Intermediate

ALGEBRA

Skills Practice Workbook
with Answers

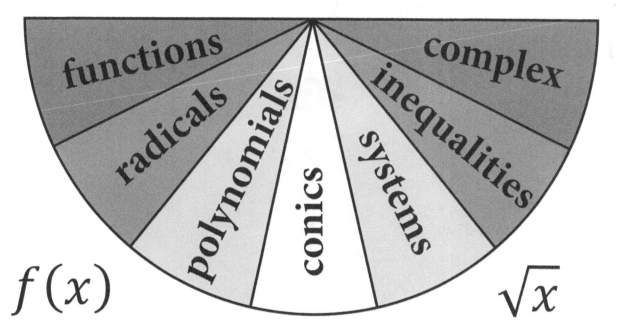

$f(x)$

functions

radicals

polynomials

conics

systems

inequalities

complex

$\sqrt{x}$

Chris McMullen, Ph.D.

Intermediate Algebra Skills Practice Workbook with Answers
Functions, Radicals, Polynomials, Conics, Systems, Inequalities, and Complex Numbers
Chris McMullen, Ph.D.

www.improveyourmathfluency.com
www.monkeyphysicsblog.wordpress.com
www.chrismcmullen.com

Zishka Publishing
ISBN: 978-1-941691-35-9

Mathematics > Algebra

CONTENTS

INTRODUCTION

The goal of this workbook is to help students develop fluency with the following intermediate algebra skills.

- properties of functions like the even function $f(t) = 2t^4 - 3t^2 + 7$
- solving for variables in radicals like $\sqrt{x^2 + 1}$ or like $x\sqrt{x}$
- simplifying irrational expressions like $\sqrt{3} - \frac{1}{\sqrt{3}}$ or like $5x^3\sqrt{80x^4}$
- dealing with fractional or negative exponents like $x^{-2/3}$
- performing calculations with complex numbers like $5 - 2i$
- finding complex conjugates
- completing the square
- solutions to the quadratic equation that involve imaginary numbers
- elimination techniques for systems of equations
- dividing polynomials, including synthetic division
- finding roots of polynomials like $p(x) = x^3 + 23x^2 + 140x + 196$
- the remainder theorem and the factor theorem
- the fundamental theorem of algebra
- equations for parabolas, hyperbolas, circles, and ellipses
- graphing functions like $y(x) = 2^x$
- working with inequalities like $-\frac{27}{x} < \frac{x}{3}$
- and more

The answer to every problem can be found in an answer key at the back of the book. Take the time to check the answers and strive to learn from any mistakes.

1 FUNCTIONS

1.1 What Is a Function?

A mathematical **function** satisfies two conditions:
- The function relates two variables by a mathematical **rule**. For any value of the independent variable, the rule indicates how to find the dependent variable.
- For each value of the independent variable, there is **exactly one** value of the dependent variable.

An example of a function is $f(x) = x^2 - 1$. In this example, the independent variable is x and the dependent variable is f. Given any value of x, the dependent variable can be found by squaring x and subtracting one. For example, when $x = 3$, we get $f(3) = 3^2 - 1 = 9 - 1 = 8$, and when $x = 4$, we get $f(4) = 4^2 - 1 = 16 - 1 = 15$.

The notation $f(x)$ is used to indicate that the **dependent** variable f is a function of the **independent** variable x. It is important to realize that $f(x)$ does NOT mean to multiply f by x. Rather, $f(x)$ means that a rule exists for how to determine f for any given value of x. The variable in parentheses is called the **argument**. The notation $f(3)$ means to evaluate the function when $x = 3$; it does NOT mean to multiply f by 3.

It is not necessary to use the symbols f and x. For example, $g(t)$ means that g is a function of the argument t and $y(x)$ means that y is a function of the argument x. Note that sometimes the function is referred to without using parentheses. For example, $f = \sqrt{x}$ is short for $f(x) = \sqrt{x}$ and $y = 5x - 2$ is short for $y(x) = 5x - 2$.

Functions can be described in different ways:
- An **equation** like $f(x) = x^2 - 5x + 6$ is one way to describe a function.
- A function can be represented by giving a **table** of values for x and f. This is common when the relationship is determined by making measurements.
- A function can be represented by giving a **graph**.
- A function can be described in **words**. For example, if a function is described as being four less than a number squared, this means that $f(x) = x^2 - 4$.

Example 1. Is f a function of x if $f = \sqrt{x+3}$?

Yes. For any value of x for which f is real, there is exactly one value of f.

Example 2. For the table below, is y a function of x? Is x a function of y?

x	y
-2	4
-1	1
0	0
1	1
2	4

y is a function of x because there is exactly one value of y for each value of x. However, x is NOT a function of y because there are multiple values of x for certain values of y. For example, when $y = 4$, one time x equals -2 and another time x equals 2. Since x has two different values when $y = 4$, x is NOT a function of y.

Example 3. For each graph below, is the variable on the vertical axis a function of the variable on the horizontal axis?

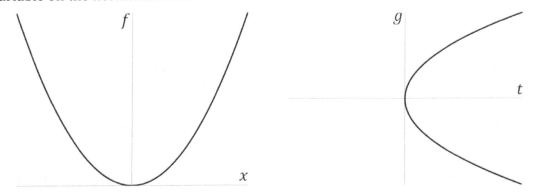

For the left graph, f is a function of x because there is exactly one value of f for each value of x. For the right graph, g is NOT a function of t because there are two values of g for each value of t (except for the point that lies at the origin). When t is positive, there are both positive and negative values for g.

Tip: The <u>vertical line test</u> can be used to determine whether or not a graph represents a function. On the left graph above, you can't draw a vertical line that intersects two points on the curve. On the right graph above, you can draw a single vertical line that intersects multiple points on the curve. The right graph fails the vertical line test.

1.1 Problems

Directions: For each equation, table, or graph, determine if f is a function of x.

(1) $f = \sqrt{x}$

(2) $f = \pm x$ (where $\pm$ means plus or minus)

(3)

x	3	6	8	2	6
f	4	9	12	5	7

(4)

x	2	6	9	8	3
f	5	8	10	9	5

(5)

(6)

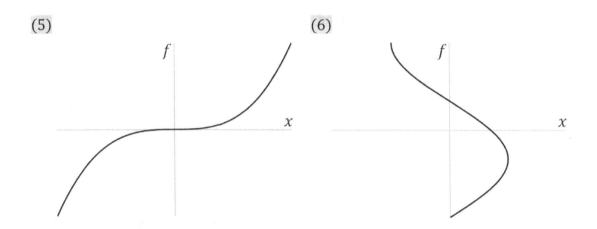

1.2 Domain and Range

The **domain** of a function refers to the possible values of the independent variable for which the function equals a real value. The **range** of a function refers to the possible values of the function. Basically, the domain refers to the possible input values while the range refers to the possible output values. For the function $f(x)$, the **domain** refers to the possible values of x while the **range** refers to the possible values of f.

Interval notation is often used to indicate the domain and range. Parentheses indicate an open interval, whereas brackets indicate a closed interval, like these examples:
- $[-2, 5]$ is the closed interval from -2 to 5, including the endpoints. If $[-2, 5]$ is the domain of $f(x)$, this means that $-2 \leq x \leq 5$. Here, x can be equal to -2, x can be equal to 5, and x can be any value in between.
- $(0, 4)$ is the open interval from 0 to 4, excluding the endpoints. If $(0, 4)$ is the domain of $f(x)$, this means that $0 < x < 4$. Here, x is greater than 0 and less than 4; x can't equal 0 or 4.
- $(3, 6]$ includes 6 and all values between 3 and 6, but doesn't include 3.
- $[-1, 1)$ includes -1 and all values between -1 and 1, but doesn't include 1.
- $(-\infty, 4) \cup (4, \infty)$ is the **union** of two intervals. It is equivalent to $x \neq 4$.
- $[0, \infty)$ is nonnegative. It includes 0 and every possible positive number. When the symbol for infinity (∞) is used, a parenthesis is always used for that limit.

The symbol $\mathbb{R}$ represents **all real numbers**. It is equivalent to $(-\infty, \infty)$.

Example 1. What are the domain and range of $f(x) = \sqrt{x}$?
The domain is $[0, \infty)$, meaning that x must be nonnegative. The reason is that only a nonnegative number has a real square root. The domain is equivalent to $0 \leq x$. The range is also $[0, \infty)$, meaning that f is nonnegative. The range is equivalent to $0 \leq f$.

Example 2. What are the domain and range of $g(t) = \frac{1}{t}$?
The domain is $t \neq 0$ since g would involve division by zero if t were zero. The domain can be expressed as $(-\infty, 0) \cup (0, \infty)$; the "$\cup$" and both intervals are needed; this is equivalent to $|t| > 0$. The range is $g \neq 0$, which is equivalent to $(-\infty, 0) \cup (0, \infty)$, since no real (and finite) value of t will make g equal zero exactly.

1.2 Problems

Directions: Indicate the domain and range of each function.

(1) $f(x) = x^2$

(2) $g(y) = \sqrt{y-2}$

(3) $y(x) = \frac{1}{x+3}$

(4) $f(t) = t^3$

(5) $z(y) = y^4 + 3y^2 + 6$

(6) $v(u) = \frac{1}{u^2}$

(7) $x(t) = \sqrt{4-t}$

(8) $g(x) = (x+5)^2$

(9) $f(y) = \frac{7}{(y+3)(y-5)}$

(10) $z(w) = 9 - w^2$

(11) $N(k) = \frac{1}{\sqrt{k}}$

(12) $p(q) = \sqrt{3q-6}$

(13) $b(a) = \frac{a}{a+1}$

(14) $Q(r) = r^{3/2}$

(15) $T(t) = \sqrt{t^2 - 9}$

(16) $h(x) = \frac{2}{x^2 + 2x - 8}$

1.3 Evaluating Functions

To **evaluate** a function at a specific value, simply replace the independent variable with the specified value in the function and simplify the answer. For example, consider the function $f(x) = x^2 - 3$. To evaluate this function at a specific value of x, simply plug the value into the expression $x^2 - 3$. The notation $f(4)$ means to evaluate $f(x)$ when x equals 4. Similarly, the notation $f(2)$ means to evaluate $f(x)$ with x equals 2. The examples below show $f(x)$ evaluated at some different values of x.

- $f(4) = 4^2 - 3 = 16 - 3 = 13$
- $f(2) = 2^2 - 3 = 4 - 3 = 1$
- $f(0) = 0^2 - 3 = 0 - 3 = -3$
- $f(-1) = (-1)^2 - 3 = 1 - 3 = -2$

Example 1. Given $f(x) = \sqrt{x + 5}$, find $f(2)$, $f(4)$, and $f(7)$.

Plug each value of x into $\sqrt{x + 5}$.

- $f(2) = \sqrt{2 + 5} = \sqrt{7}$. This irrational number can't be simplified further.
- $f(4) = \sqrt{4 + 5} = \sqrt{9} = 3$. Only the positive root applies because a function is necessarily single-valued (Sec. 1.1). Note: The single-valued nature of a function is important in higher-level mathematics (especially, in calculus).
- $f(7) = \sqrt{7 + 5} = \sqrt{12} = \sqrt{(4)(3)} = \sqrt{4}\sqrt{3} = 2\sqrt{3}$. We factored out the perfect square in order to express the answer in standard form. We applied the rule $\sqrt{xy} = \sqrt{x}\sqrt{y}$. Although $\sqrt{12}$ is a correct answer, it isn't in standard form. The best answer is $2\sqrt{3}$, which is equivalent to $\sqrt{12}$.

Example 2. Given $g(t) = \frac{t-1}{t+1}$, find $g(2)$ and $g(3)$.

Plug each value of x into $\frac{t-1}{t+1}$.

- $g(2) = \frac{2-1}{2+1} = \frac{1}{3}$. This reduced fraction can't be simplified further.
- $g(3) = \frac{3-1}{3+1} = \frac{2}{4} = \frac{1}{2}$. We divided the numerator and denominator each by 2 in order to reduce $\frac{2}{4}$ to $\frac{1}{2}$.

1.3 Problems

Directions: Evaluate each function at the specified values.

(1) Given $f(x) = x^2 - 3x + 8$, find $f(3)$, $f(2)$, $f(1)$, $f(0)$, and $f(-1)$.

(2) Given $g(y) = 7y - 5$, find $g(2)$, $g(1)$, $g\left(\frac{1}{2}\right)$, $g\left(-\frac{1}{2}\right)$, and $g(-1)$.

(3) Given $y(x) = \sqrt{25 - x^2}$, find $y(5)$, $y(4)$, $y(3)$, $y(-1)$, and $y(-3)$.

(4) Given $v(t) = \frac{t}{3} - \frac{4}{t}$, find $v(4)$, $v(3)$, $v(2)$, $v(1)$, and $v(-12)$.

(5) Given $f(x) = x^{2/3}$, find $f(27)$, $f(8)$, $f(-8)$, and $f(-27)$.

1.4 Multivariable Functions

Not all functions have a single argument like $g(x) = \sqrt{x}$ or $h(y) = \frac{2}{y}$. Some functions have two or more arguments, like $f(x, y) = x^2 + y^2$ or $p(x, y, z) = 3x^3y^2z$. A multivariable function can be evaluated by plugging in the values for each variable.

Example 1. Given $f(x, y) = x^3 - 3y$, find $f(2, 4)$.
Plug $x = 2$ and $y = 4$ into the function: $f(2, 4) = 2^3 - 3(4) = 8 - 12 = -4$.

1.4 Problems

Directions: Evaluate each function at the specified values.

(1) Given $f(x, y) = y^2 - x^2$, find $f(3, 2)$.

(2) Given $g(x, t) = \sqrt{xt}$, find $g(6, 54)$.

(3) Given $Q(p, r) = \frac{1}{p} + \frac{p}{r}$, find $Q(4, 3)$.

(4) Given $d(a, b, c) = c^2 - ab$, find $d(3, 4, 5)$.

1.5 Step Functions and Piecewise Functions

A simple example of a **step function** is:

$$f(x) = \begin{cases} 0 & \text{if } x \le 0 \\ 1 & \text{if } x > 0 \end{cases}$$

This step function equals one of two values (0 or 1), depending on the value of the argument. For negative values of x or when x equals zero, the function equals 0, while for positive values of x, the function equals 1. The step function is an example of a **piecewise function**. A piecewise function equals different expressions in different sections of its domain. Another example of a piecewise function is:

$$g(t) = \begin{cases} 1 & \text{if } & t < -1 \\ t & \text{if } & -1 < t < 1 \\ t^2 & \text{if } & 1 < t \end{cases}$$

Note that the domain of $f(x)$ above is $\mathbb{R}$, whereas the domain of $g(t)$ above excludes the points $x = \pm 1$ because the definition of $g(t)$ uses less than ($<$) signs, but does not use less than or equal to signs ($\le$).

A function with **absolute values** can be expressed as a piecewise function. For example, consider the function $f(x) = |x|$. The straight lines mean to take the absolute value of x (which means to ignore the sign of x). For example, $f(-2) = |-2| = 2$. When x is nonnegative, the absolute values have no effect. For example, $f(3) = |3| = 3$. This function is equivalent to the following piecewise function:

$$f(x) = \begin{cases} -x & \text{if } x < 0 \\ x & \text{if } x \ge 0 \end{cases}$$

When x is negative, $-x$ equals a positive number, and when x is positive, x also equals a positive number.

Example 1. Find $f(-2)$, $f(0)$, and $f(3)$ for the function below.

$$f(x) = \begin{cases} x^3 - 1 & \text{if } x \le 0 \\ x^2 + 1 & \text{if } x > 0 \end{cases}$$

When x is negative or zero, use $x^3 - 1$, and when x is positive, use $x^2 + 1$: $f(-2) = (-2)^3 - 1 = -8 - 1 = -9$, $f(0) = 0^3 - 1 = -1$, and $f(3) = 3^2 + 1 = 9 + 1 = 10$.

1.5 Problems

Directions: Evaluate each function at the specified values.

(1) For the function below, find $f(-3)$, $f(0)$, $f(2)$, $f(3)$, and $f(5)$.
$$f(x) = \begin{cases} x^2 - 9 & \text{if} \quad x \leq 3 \\ 3 - x & \text{if} \quad x > 3 \end{cases}$$

(2) For the function below, find $g\left(-\frac{1}{4}\right)$, $g(0)$, $g(1)$, $g\left(\frac{3}{2}\right)$, and $g(2)$.
$$g(t) = \begin{cases} 1 - t & \text{if} & t < 0 \\ 2t + 1 & \text{if} & 0 \leq t < 2 \\ t^2 & \text{if} & 2 \leq t \end{cases}$$

(3) For the function $f(x) = |3 - 2x|$, find $f(-3)$, $f(0)$, $f(1)$, $f(2)$, and $f(3)$.

(4) For the function $g(x) = |2 - x^2|$, find $g(-3)$, $g(-1)$, $g(0)$, $g(1)$, and $g(3)$.

(5) For the function $h(x) = |x - \sqrt{x}|$, find $h(0)$, $h\left(\frac{1}{4}\right)$, $h(1)$, $h\left(\frac{9}{4}\right)$, and $h(4)$.

1.6 Composite Functions

The notation $f\big(g(x)\big)$ indicates that f is a **composite function**. To find $f\big(g(x)\big)$, first plug x into $g(x)$, and then plug $g(x)$ in as the argument for $f(x)$. The notation $f \circ g$ is an alternative way to write $f\big(g(x)\big)$. In order for the composite function $f\big(g(x)\big)$ to make sense, the range of $g(x)$ must lie in the domain of f.

Example 1. Given $f(x) = x + 8$ and $g(x) = \sqrt{4x + 5}$, find $f\big(g(11)\big)$.

Evaluate $g(x)$ at $x = 11$ to get $g(11) = \sqrt{4(11) + 5} = \sqrt{44 + 5} = \sqrt{49} = 7$. Now use $g(11) = 7$ as the argument for $f(x)$. This means to replace x with 7 in $f(x)$.
$$f\big(g(11)\big) = f(7) = 7 + 8 = 15$$

Example 2. Given $f(x) = 2x^2 - 1$ and $g(x) = 3x - 2$, find $f\big(g(x)\big)$.

Use $3x - 2$ as the argument of f. This means to replace x with $3x - 2$ in $2x^2 - 1$.
$$f\big(g(x)\big) = f(3x - 2) = 2(3x - 2)^2 - 1 = 2(9x^2 - 12x + 4) - 1$$
$$f\big(g(x)\big) = 18x^2 - 24x + 8 - 1 = 18x^2 - 24x + 7$$

Example 3. Given $f(x) = 9 - 4x$, find $f\big(f(x)\big)$.

Use $9 - 4x$ as the argument of f. This means to replace x with $9 - 4x$ in $9 - 4x$.
$$f\big(f(x)\big) = f(9 - 4x) = 9 - 4(9 - 4x) = 9 - 4(9) - 4(-4x)$$
$$f\big(f(x)\big) = 9 - 36 + 16x = -27 + 16x = 16x - 27$$

1.6 Problems

Directions: Evaluate each composite function.

(1) Given $f(x) = 5x - 2$ and $g(x) = 4x + 3$, find $f\big(g(6)\big)$, $g\big(f(6)\big)$, $f\big(g(x)\big)$, and $g\big(f(x)\big)$.

(2) Given $f(t) = \frac{24}{t}$ and $g(t) = t^2 + 3$, find $f(g(1))$, $f(g(3))$, and $f(g(t))$.

(3) Given $f(x) = \sqrt{7x + 2}$, find $f(f(1))$, $f(f(2))$, $f(f(14))$, and $f(f(x))$.

(4) Given $f(x) = x^2 - 4$ and $g(x) = x + 8$, find $f(f(3))$, $f(g(3))$, $g(f(3))$, and $g(g(3))$.

(5) Given $p(y) = -8 - y^2$ and $q(y) = \frac{12}{y}$, find $p(q(4))$ and $q(p(4))$.

(6) Given $f(x) = 3x + 2$, $g(x) = 4x - 5$, and $h(x) = 4x + 3$, find $f\left(g(h(x))\right)$.

1.7 Inverse Functions

Two functions, $f(x)$ and $g(x)$, are **inverse** functions of one another if:
- $f(g(x)) = x$ for every value of x in the domain of $g(x)$.
- $g(f(x)) = x$ for every value of x in the domain of $f(x)$.

Basically, a function is "undone" by its inverse function, meaning that when the inverse function acts on the function, the two effects essentially cancel out, resulting in x. The notation $f^{-1}(x)$ indicates that f^{-1} is the **inverse** of f. **Caution**: The superscript of -1 in f^{-1} does NOT mean to find $\frac{1}{f}$. Instead, it means to find the inverse of f. See below.

Not all functions have an inverse. If two different inputs both result in the same output, the function does NOT have an inverse. For example, the function $f(x) = x^2$ does not have an inverse because $f(-2) = (-2)^2 = 4$ equals $f(2) = 2^2 = 4$. That is, the inputs of -2 and 2 both result in the same output of 4. (We would need to restrict the domain of this function to $x \geq 0$ or to $x \leq 0$ in order for its inverse to make sense.)

Suppose that we know the function $f(x)$ and wish to find its inverse function, $f^{-1}(x)$. One way to do this to use f^{-1} as the argument of $f(x)$, making the composite function $f(f^{-1})$. By the definition of an inverse function, $f(f^{-1}) = x$. Isolate f^{-1} in this equation to determine the inverse function. This technique is illustrated in the first example.

Example 1. Find the inverse of $f(x) = 2x + 3$.
Use f^{-1} as the argument of $f(x)$ and set this equal to x because $f(f^{-1}) = x$.
$$2f^{-1} + 3 = x$$
Now isolate f^{-1}. Subtract 3 from both sides: $2f^{-1} = x - 3$. Divide by 2 on both sides:
$$f^{-1}(x) = \frac{x - 3}{2}$$
Check: $f(f^{-1}) = f\left(\frac{x-3}{2}\right) = 2\left(\frac{x-3}{2}\right) + 3 = x - 3 + 3 = x$.

Example 2. Does $f(x) = (x - 3)(x - 4)$ have an inverse?
No. There are two different values of x that produce the same value for $f(x)$ because $f(3) = 0$ and $f(4) = 0$. When the input is 3 or 4, the output is the same value (zero). Although $f(x)$ satisfies the definition of a function, its inverse would not.

1.7 Problems

Directions: Does each function have an inverse? If so, find it. If not, explain why.

(1) $f(x) = 5x - 4$

(2) $g(x) = 8 - 7x$

(3) $h(t) = t^4$

(4) $p(y) = y^5$

(5) $f(x) = \dfrac{6}{x}$

(6) $q(t) = \dfrac{9}{2t+5}$

(7) $g(x) = \dfrac{x+2}{x-3}$

(8) $h(k) = \dfrac{4k-7}{5k+8}$

(9) $f(y) = \sqrt{3y + 8}$

(10) $f(x) = x^2 - 4x - 12$

1.8 Even Functions and Odd Functions

To determine whether a function is even, odd, or neither, evaluate the function at $-x$ and compare $f(-x)$ with $f(x)$:

- If $f(-x) = f(x)$ for every value of x in f's domain, the function is **even**.
- If $f(-x) = -f(x)$ for every value of x in f's domain, the function is **odd**.
- If neither statement above is true, the function is neither even nor odd.

For example, $f(x) = x^2$ is an even function because $f(-x) = (-x)^2 = x^2$ is the same as $f(x)$ for any real value of x, whereas $g(x) = x^3$ is an odd function because $g(-x) = (-x)^3 = -x^3$ is the negative of $g(x)$.

If every term of a **polynomial** has an even power of x or a constant, like $3x^4 - 2x^2 + 5$, the polynomial is an **even** function. If every term of a polynomial has an odd power of x (and no term is a constant), like $x^5 + 4x^3 - 7x$, the polynomial is an **odd** function. If a polynomial has both even and odd terms, like $x^2 - 6x$ (or has odd terms and a constant term, like $3x^3 + 8$), the polynomial is neither even nor odd.

Chapter 7 explores the graphs of even and odd functions.

Example 1. Is $f(x) = 4x^2 - 5$ even, odd, or neither?

It is even because $f(-x) = 4(-x)^2 - 5 = 4x^2 - 5 = f(x)$. Since every term of the polynomial has an even power (or is a constant term), the polynomial is even.

Example 2. Is $g(x) = x^5 - 7x^3 + 9$ even, odd, or neither?

It is neither because $g(-x) = (-x)^5 - 7(-x)^3 + 9 = -x^5 + 7x^3 + 9$ is neither equal to $g(x)$ nor is <u>every</u> term negated. When a polynomial contains both even powers and odd powers (or both odd powers and a constant term, as is the case here), it is neither odd nor even.

Example 3. Is $f(t) = \frac{2}{t}$ even, odd, or neither?

It is odd because $f(-t) = \frac{2}{-t} = -\frac{2}{t} = -f(t)$.

1.8 Problems

Directions: Indicate whether each function is even, odd, or neither.

(1) $f(x) = x^5$

(2) $g(x) = x^4$

(3) $h(t) = t^8 - 5t^6 - 3$

(4) $p(y) = -4y^9 + 2y^5 - y$

(5) $q(r) = 3r + 8$

(6) $f(x) = \sqrt{x}$

(7) $g(y) = \sqrt{y^2 - 9}$

(8) $v(t) = \frac{5}{t^3 - 6t}$

(9) $f(x) = (x + 7)(x - 7)$

(10) $q(z) = (z - 3)(z + 6)$

(11) $f(y) = \sqrt[3]{y}$

(12) $y(x) = 4^x$

2 RADICALS

2.1 Powers of Square Roots

When $\sqrt{x}$ is multiplied by itself, the effect is to remove the radical: $\sqrt{x}\sqrt{x} = x$. When $\sqrt{x}$ is squared, the result is the same: $\left(\sqrt{x}\right)^2 = x$. If there is an expression inside of the radical and the radical is squared, the radical is similarly removed: $\sqrt{3x+2}\sqrt{3x+2} = 3x + 2$. If a square root is raised to an **even power**, remove the radical and raise the expression inside to one-half of the power. For example, $\left(\sqrt{2x}\right)^6 = (2x)^{6/2} = (2x)^3 = 2^3 x^3 = 8x^3$. The reason is that a square root is equivalent to a power of one-half: $\sqrt{x} = x^{1/2}$. Set $m = \frac{1}{2}$ in the rule $(x^m)^n = x^{mn}$ to see that $\left(\sqrt{x}\right)^n = \left(x^{1/2}\right)^n = x^{n/2}$. Note: The rule $(ax)^n = a^n x^n$ explains why $(2x)^3 = 2^3 x^3$.

If a square root is raised to an **odd power**, apply the rule $x^n = x^{n-1}x$ to rewrite the radical raised to an even power. For example, $\left(\sqrt{x}\right)^5 = \left(\sqrt{x}\right)^4\sqrt{x} = x^{4/2}\sqrt{x} = x^2\sqrt{x}$.

Example 1. $\left(\sqrt{x-7}\right)^2 = x - 7$ **Example 2.** $\sqrt{5x}\sqrt{5x} = 5x$

Example 3. $\left(\sqrt{3x}\right)^8 = (3x)^{8/2} = (3x)^4 = 3^4 x^4 = 81x^4$

Example 4. $\left(\sqrt{x+2}\right)^5 = \left(\sqrt{x+2}\right)^4\sqrt{x+2} = (x+2)^2\sqrt{x+2} = (x^2 + 4x + 4)\sqrt{x+2}$

2.1 Problems

Directions: Simplify each expression.

(1) $\left(\sqrt{9x}\right)^2$ (2) $\sqrt{3}\sqrt{3}$

(3) $\sqrt{5x-4}\sqrt{5x-4}$ (4) $\left(\sqrt{x+6}\right)^2$

---

(5) $\left(\sqrt{5x}\right)^4$

(6) $\left(\sqrt{3x}\right)^3$

(7) $\left(\sqrt{x-9}\right)^8$

(8) $\left(\sqrt{2x+7}\right)^5$

(9) $\left(\sqrt{5}\right)^4\left(\sqrt{10x}\right)^3$

(10) $\left(\sqrt{x+6}\right)^4\left(\sqrt{x-6}\right)^5$

(11) $\sqrt{x}\sqrt{x}\sqrt{x}\sqrt{x}$

(12) $\sqrt{x+2}\sqrt{x+2}\sqrt{x-2}\sqrt{x-2}$

(13) $\sqrt{x-4}\sqrt{2x+1}\sqrt{x-4}\sqrt{2x+1}$

(14) $\left(\sqrt{3x}\right)^4\left(\sqrt{9x}\right)^2$

(15) $\left(\sqrt{8x}\right)^3\sqrt{x}$

(16) $\left(\sqrt{7x}\right)^5\left(\sqrt{2x}\right)^3$

(17) $\left(\sqrt{2x}\right)^6\left(\sqrt{6x}\right)^4\left(\sqrt{3x}\right)^2$

(18) $\left(\sqrt{3x}\right)^5\left(\sqrt{6x}\right)^3\left(\sqrt{2x}\right)$

(19) $\left(\sqrt{5x}\right)^4\left(\sqrt{x}\right)^3\left(\sqrt{7x}\right)^2$

(20) $\left(\sqrt{6x}\right)^5\left(\sqrt{7x}\right)^4\left(\sqrt{2x}\right)^3$

2.2 Factor Perfect Squares

The rule $\sqrt{uv} = \sqrt{u}\sqrt{v}$ allows a perfect square to be factored out of a radical. As an example, the perfect square 36 can be factored out of $\sqrt{108}$ because $108 = (36)(3)$:

$$\sqrt{108} = \sqrt{(36)(3)} = \sqrt{36}\sqrt{3} = 6\sqrt{3}$$

A squared variable may similarly be factored out of a radical.

$$\sqrt{2x^2} = \sqrt{2}\sqrt{x^2} = x\sqrt{2}$$

If a variable is raised to an even power, its power is cut in half when it is factored out of a radical. For example, $\sqrt{3x^{12}} = x^6\sqrt{3}$. If a variable is raised to an odd power, apply the rule $x^n = x^{n-1}x$ to rewrite it in terms of an even power. For example, $\sqrt{7x^9} = \sqrt{7x^8x} = \sqrt{x^8}\sqrt{7x} = x^4\sqrt{7x}$.

Example 1. $\sqrt{18} = \sqrt{(9)(2)} = \sqrt{9}\sqrt{2} = 3\sqrt{2}$

Example 2. $\sqrt{75x^6} = \sqrt{(25)(3)}\sqrt{x^6} = x^3\sqrt{25}\sqrt{3} = 5x^3\sqrt{3}$

Example 3. $\sqrt{80x^7} = \sqrt{(16)(5)x^6x} = \sqrt{16}\sqrt{x^6}\sqrt{5x} = 4x^3\sqrt{5x}$

2.2 Problems

Directions: Factor the perfect squares out of the radical.

(1) $\sqrt{50}$

(2) $\sqrt{45x^4}$

(3) $\sqrt{12x^7}$

(4) $\sqrt{98x^8}$

(5) $\sqrt{63x^2}$

(6) $\sqrt{192x^{11}}$

(7) $\sqrt{1100x^5}$

(8) $\sqrt{288x^{16}}$

(9) $\sqrt{567x^6}$

(10) $\sqrt{216x^9}$

2.3 Factor Perfect Roots

Perfect cubes may be factored out of cube roots similar to how perfect squares may be factored out of square roots. For example, $\sqrt[3]{54x^6} = \sqrt[3]{(27)(2)}\sqrt[3]{x^6} = \sqrt[3]{27}\sqrt[3]{2}\sqrt[3]{x^6}$ $= 3x^2\sqrt[3]{2}$ because $(x^2)^3 = x^6$. Higher roots work the same way. For example, $\sqrt[5]{96x^5}$ $= \sqrt[5]{(32)(3)}\sqrt[5]{x^5} = \sqrt[5]{32}\sqrt[5]{3}\sqrt[5]{x^5} = 2x\sqrt[5]{3}$. For an odd root, a minus sign inside of the root may be factored out (whereas for an even root, a minus sign inside of the root would result in an imaginary number). For example, $\sqrt[3]{-x^4} = \sqrt[3]{-x^3x} = \sqrt[3]{-x^3}\sqrt[3]{x} =$ $-x\sqrt[3]{x}$. If the variable doesn't factor perfectly, rewrite it in terms of a lower power that does. For example, $\sqrt[4]{81x^7} = \sqrt[4]{81}\sqrt[4]{x^4}\sqrt[4]{x^3} = 3x\sqrt[4]{x^3}$.

Example 1. $\sqrt[4]{48x^{15}} = \sqrt[4]{(16)(3)x^{12}x^3} = \sqrt[4]{16}\sqrt[4]{x^{12}}\sqrt[4]{3x^3} = 2x^3\sqrt[4]{3x^3}$

Example 2. $\sqrt[3]{-135x^8} = \sqrt[3]{(-27)(5)x^6x^2} = \sqrt[3]{-27}\sqrt[3]{x^6}\sqrt[3]{5x^2} = -3x^2\sqrt[3]{5x^2}$

2.3 Problems

Directions: Factor the perfect powers out of the radical.

(1) $\sqrt[4]{162x^8}$

(2) $\sqrt[3]{24x^{12}}$

(3) $\sqrt[5]{160x^{14}}$

(4) $\sqrt[4]{80x^7}$

(5) $\sqrt[3]{250x^{10}}$

(6) $\sqrt[6]{192x^{27}}$

(7) $\sqrt[4]{512x^9}$

(8) $\sqrt[3]{-81x^5}$

(9) $\sqrt[5]{-243x^{11}}$

(10) $\sqrt[4]{70{,}000x^{18}}$

2.4 Rationalize the Denominator

Many algebra instructors prefer for students to express their answers in standard form, which means to reduce any fractions, factor perfect squares out of square roots, and rationalize the denominator. A fraction with a radical in the denominator like $\frac{6}{\sqrt{2x}}$ has an irrational denominator. To **<u>rationalize the denominator</u>** of $\frac{6}{\sqrt{2x}}$, multiply by $\frac{\sqrt{2x}}{\sqrt{2x}}$:

$$\frac{6}{\sqrt{2x}} = \frac{6}{\sqrt{2x}}\frac{\sqrt{2x}}{\sqrt{2x}} = \frac{6\sqrt{2x}}{2x} = \frac{3\sqrt{2x}}{x}$$

Example 1. $\frac{x}{\sqrt{5}} = \frac{x}{\sqrt{5}}\frac{\sqrt{5}}{\sqrt{5}} = \frac{x\sqrt{5}}{5}$ **Example 2.** $\frac{12}{\sqrt{3x}} = \frac{12}{\sqrt{3x}}\frac{\sqrt{3x}}{\sqrt{3x}} = \frac{12\sqrt{3x}}{3x} = \frac{4\sqrt{3x}}{x}$

2.4 Problems

Directions: Rationalize the denominator of each fraction.

(1) $\frac{1}{\sqrt{2}}$

(2) $\frac{6}{\sqrt{3}}$

(3) $\frac{x}{2\sqrt{7}}$

(4) $\frac{24x}{\sqrt{6}}$

(5) $\frac{1}{\sqrt{x}}$

(6) $\frac{5}{\sqrt{10x}}$

(7) $\frac{9x}{\sqrt{3x}}$

(8) $\frac{1}{x\sqrt{x}}$

(9) $\frac{\sqrt{3x}}{\sqrt{6}}$

(10) $\frac{\sqrt{5}}{\sqrt{15x}}$

(11) $\frac{x^2}{2\sqrt{x}}$

(12) $\frac{\sqrt{35}}{\sqrt{21}}$

2.5 Multiply by the Conjugate

The expressions $3 - \sqrt{2}$ and $3 + \sqrt{2}$ are **conjugate** expressions in the sense that the product of these expressions doesn't have an irrational part:

$$\left(3 - \sqrt{2}\right)\left(3 + \sqrt{2}\right) = (3)(3) + 3\sqrt{2} - 3\sqrt{2} - \sqrt{2}\sqrt{2} = 9 - 2 = 7$$

A denominator like $\frac{4-\sqrt{3}}{4+\sqrt{3}}$ can be rationalized by multiplying both the numerator and the denominator by the conjugate of the denominator.

$$\frac{4 - \sqrt{3}}{4 + \sqrt{3}} = \frac{4 - \sqrt{3}}{4 + \sqrt{3}}\left(\frac{4 - \sqrt{3}}{4 - \sqrt{3}}\right) = \frac{4(4) - 4\sqrt{3} - 4\sqrt{3} + \sqrt{3}\sqrt{3}}{4(4) - 4\sqrt{3} + 4\sqrt{3} - \sqrt{3}\sqrt{3}} = \frac{16 - 8\sqrt{3} + 3}{16 - 3} = \frac{19 - 8\sqrt{3}}{13}$$

Example 1. $\frac{3-\sqrt{2}}{3+\sqrt{5}} = \frac{3-\sqrt{2}}{3+\sqrt{5}}\left(\frac{3-\sqrt{5}}{3-\sqrt{5}}\right) = \frac{3(3)-3\sqrt{5}-3\sqrt{2}+\sqrt{2}\sqrt{5}}{3(3)-3\sqrt{5}+3\sqrt{5}-\sqrt{5}\sqrt{5}} = \frac{9-3\sqrt{5}-3\sqrt{2}+\sqrt{10}}{9-5} = \frac{9-3\sqrt{5}-3\sqrt{2}+\sqrt{10}}{4}$

Example 2. $\frac{\sqrt{6}}{5-\sqrt{2}} = \frac{\sqrt{6}}{5-\sqrt{2}}\left(\frac{5+\sqrt{2}}{5+\sqrt{2}}\right) = \frac{5\sqrt{6}+\sqrt{6}\sqrt{2}}{5(5)+5\sqrt{2}-5\sqrt{2}-\sqrt{2}\sqrt{2}} = \frac{5\sqrt{6}+\sqrt{12}}{25-2} = \frac{5\sqrt{6}+\sqrt{4}\sqrt{3}}{23} = \frac{5\sqrt{6}+2\sqrt{3}}{23}$

2.5 Problems

Directions: Rationalize the denominator of each fraction.

(1) $\frac{2+\sqrt{3}}{2-\sqrt{3}}$

(2) $\frac{\sqrt{10}}{2+\sqrt{5}}$

(3) $\frac{3-\sqrt{7}}{4-\sqrt{7}}$

(4) $\dfrac{3+\sqrt{6}}{9+\sqrt{2}}$

(5) $\dfrac{\sqrt{3}+\sqrt{2}}{\sqrt{3}-\sqrt{2}}$

(6) $\dfrac{x-\sqrt{x}}{x+\sqrt{x}}$

(7) $\dfrac{1+\sqrt{x}}{2-\sqrt{x}}$

(8) $\dfrac{x-\sqrt{6}}{x+\sqrt{3}}$

(9) $\dfrac{\sqrt{x}}{\sqrt{x}-1}$

(10) $\dfrac{\sqrt{x}+\sqrt{3}}{\sqrt{x}-\sqrt{2}}$

(11) $\dfrac{x+3}{3+2\sqrt{x}}$

2.6 Fractional Exponents

An exponent of the form $x^{1/n}$ represents the nth root: $x^{1/n} = \sqrt[n]{x}$. For example, $8^{1/3} = \sqrt[3]{8} = 2$ and $81^{1/4} = \sqrt[4]{81} = 3$. Technically, an even root has both positive and negative solutions. For example, $4^{1/2} = \sqrt{4} = \pm 2$ because $2^2 = 4$ and $(-2)^2 = 4$. However, in the context of algebra, usually only the positive root is desired because functions are single-valued (recall Chapter 1). Yet there is an important exception: When you need to take a root of both sides of an equation to solve for an unknown, in that case you must consider all possible roots in order to find every solution to the equation (Sec.'s 2.9 and 2.11). For an odd root, the answer has the same sign as what is inside the radical. For example, $(-1000)^{1/3} = \sqrt[3]{-1000} = -10$ because $(-10)^3 = -1000$.

Use the rule $(x^a)^b = x^{ab}$ to evaluate a fractional exponent: $x^{m/n} = \left(x^{1/n}\right)^m$. For example, $4^{5/2} = \left(4^{1/2}\right)^5 = \left(\sqrt{4}\right)^5 = 2^5 = 32$ and $125^{2/3} = \left(125^{1/3}\right)^2 = \left(\sqrt[3]{125}\right)^2 = 5^2 = 25$.

A negative exponent involves a reciprocal: $x^{-a} = \frac{1}{x^a}$ and $x^{-m/n} = \left(\frac{1}{x}\right)^{m/n}$. For example, $8^{-4/3} = \left(\frac{1}{8}\right)^{4/3} = \left(\frac{1}{8^{1/3}}\right)^4 = \left(\frac{1}{\sqrt[3]{8}}\right)^4 = \left(\frac{1}{2}\right)^4 = \frac{1}{16}$ and $9^{-1/2} = \frac{1}{9^{1/2}} = \frac{1}{\sqrt{9}} = \frac{1}{3}$.

Example 1. $625^{1/4} = \sqrt[4]{625} = 5$ because $5^4 = (5)(5)(5)(5) = (25)(25) = 625$

Example 2. $(27x^6)^{4/3} = (27)^{4/3}(x^6)^{4/3} = \left(27^{1/3}\right)^4\left(x^{6/3}\right)^4 = \left(\sqrt[3]{27}\right)^4(x^2)^4 = 3^4 x^8 = 81x^8$

Example 3. $(7x^3)^{-2} = \frac{1}{(7x^3)^2} = \frac{1}{49x^6}$ $\qquad$ **Example 4.** $(-8)^{-1/3} = \frac{1}{(-8)^{1/3}} = \frac{1}{\sqrt[3]{-8}} = -\frac{1}{2}$

Example 5. $\left(\frac{8}{27}\right)^{-1/3} = \left(\frac{27}{8}\right)^{1/3} = \frac{\sqrt[3]{27}}{\sqrt[3]{8}} = \frac{3}{2}$ $\quad$ **Example 6.** $\left(\frac{1}{100}\right)^{-1/2} = 100^{1/2} = \sqrt{100} = 10$

Example 7. $(16x^4)^{-3/4} = \left(\frac{1}{16x^4}\right)^{3/4} = \left(\frac{1}{16^{1/4}x^{4/4}}\right)^3 = \left(\frac{1}{x\sqrt[4]{16}}\right)^3 = \left(\frac{1}{2x}\right)^3 = \frac{1}{8x^3}$

Example 8. $\left(\frac{8x}{125}\right)^{-2/3} = \left(\frac{125}{8x}\right)^{2/3} = \left(\frac{125^{1/3}}{8^{1/3}x^{1/3}}\right)^2 = \left(\frac{\sqrt[3]{125}}{\sqrt[3]{8}x^{1/3}}\right)^2 = \left(\frac{5}{2x^{1/3}}\right)^2 = \frac{25}{4x^{2/3}}$

2.6 Problems

Directions: Simplify each expression.

(1) $125^{1/3}$

(2) $36^{1/2}$

(3) $(-32)^{1/5}$

(4) $(256x^{12})^{1/4}$

(5) $16^{3/2}$

(6) $27^{2/3}$

(7) $81^{3/4}$

(8) $(-125)^{5/3}$

(9) $(4x^6)^{5/2}$

(10) $(32x^5)^{2/5}$

(11) 10^{-2}

(12) $\left(4x^{2/3}\right)^{-3}$

(13) $256^{-1/4}$

(14) $(-27x)^{-1/3}$

(15) $\left(\frac{125}{64}\right)^{-2/3}$

(16) $\left(\frac{16x^6}{81}\right)^{-3/4}$

(17) $\left(\frac{1}{64}\right)^{-1/2}$

(18) $\left(\frac{1}{81x^8}\right)^{-1/4}$

(19) $\left(\frac{27x^{9/2}}{64}\right)^{-4/3}$

(20) $\left(\frac{32x^{-10}}{243}\right)^{-2/5}$

2.7 Distribute into Radicals

An expression gets squared when it is distributed into a square root. For example, $2x\sqrt{3x^3 - 4} = \sqrt{(2x)^2(3x^3 - 4)} = \sqrt{4x^2(3x^3 - 4)} = \sqrt{12x^5 - 16x^2}$. The reason for this is that $2x = \sqrt{(2x)^2}$. Similarly, when distributing an expression into an nth root, raise it to the power of n. For example,

$$4x^2 \sqrt[3]{2x^2 + 3x} = \sqrt[3]{(4x^2)^3(2x^2 + 3x)} = \sqrt[3]{64x^6(2x^2 + 3x)} = \sqrt[3]{128x^8 + 192x^7}$$

The reason is that $4x^2 = \sqrt[3]{(4x^2)^3}$. To distribute an expression into a fractional power, raise it to the reciprocal of the exponent. For example,

$$9x(5x^2 + 3)^{2/3} = \left[(9x)^{3/2}(5x^2 + 3)\right]^{2/3} = \left[9^{3/2}x^{3/2}(5x^2 + 3)\right]^{2/3}$$
$$= \left[27x^{3/2}(5x^2 + 3)\right]^{2/3} = \left(135x^{7/2} + 81x^{3/2}\right)^{2/3}$$

In the last step, $x^{3/2}x^2 = x^{3/2+2} = x^{7/2}$ according to the rule $x^a x^b = x^{a+b}$. Note that $\left[(9x)^{2/3}\right]^{3/2} = 9x$ according to rule $(x^a)^b = x^{ab}$ since $\left(\frac{2}{3}\right)\left(\frac{3}{2}\right) = 1$. Also, $9^{3/2} = \left(\sqrt{9}\right)^3$.

Example 1. $10x^3 \sqrt{2x^5 + 3x^3} = \sqrt{(10x^3)^2(2x^5 + 3x^3)} = \sqrt{100x^6(2x^5 + 3x^3)} = \sqrt{200x^{11} + 300x^9}$

Example 2. $4x \sqrt[3]{3x^4 + 5x^2} = \sqrt[3]{(4x)^3(3x^4 + 5x^2)} = \sqrt[3]{64x^3(3x^4 + 5x^2)} = \sqrt[3]{192x^7 + 320x^5}$

Example 3. $8x^5(7x - 2)^{3/4} = \left[8^{4/3}x^{20/3}(7x - 2)\right]^{3/4} = \left(112x^{23/3} - 32x^{20/3}\right)^{3/4}$ Note: $8^{4/3} = \left(\sqrt[3]{8}\right)^4 = 16$.

2.7 Problems

Directions: Distribute each expression into the radical.

(1) $6x \sqrt{3x^3 + 5x}$

(2) $5x^2 \sqrt{7x - 6}$

(3) $3x \sqrt[3]{x^4 + 2x^2 + 5}$

(4) $2x^2 \sqrt[5]{5x - 9}$

(5) $8x^3 \sqrt{5x^7 - 3x^4 + x}$

(6) $7x^4 \sqrt[3]{2x^3 + 7x}$

(7) $2x \sqrt[4]{x^4 - 8x^2 - 4}$

(8) $10x^3 \sqrt{15x^5 - 6x^3 + 9x}$

(9) $3x(x^2 + 4x)^{1/3}$

(10) $5x^2(3x - 2)^{1/4}$

(11) $4x^4(5x^2 + 6)^{2/3}$

(12) $32x^5(4x^3 - 8x)^{5/4}$

(13) $64x^{12}(x^2 + 3x - 7)^{3/5}$

(14) $125x^6(9x^3 - 6)^{3/2}$

(15) $256x^{24}(5x^4 + 8x)^{4/3}$

(16) $243x^{10}(2x^8 - 3x^6 + 6x^4)^{-5/2}$

2.8 Irrational Expressions

When multiplying irrational expressions, note that $\sqrt{x}\sqrt{x} = x$ and $\sqrt{x}\sqrt{y} = \sqrt{xy}$. When dividing irrational expressions, note that $\frac{\sqrt{x}}{\sqrt{x}} = 1$ and $\frac{\sqrt{x}}{\sqrt{y}} = \sqrt{\frac{x}{y}}$. Express the final answer in standard form by rationalizing the denominator (Sec.'s 2.4-2.5), factoring out any perfect squares (Sec. 2.2), and reducing any fractions.

Example 1. $4x^2 + 5\sqrt{x} + 3x^2 - 2\sqrt{x} = (4 + 3)x^2 + (5 - 2)\sqrt{x} = 7x^2 + 3\sqrt{x}$
Note: $4x^2$ and $3x^2$ are like terms, and $5\sqrt{x}$ and $-2\sqrt{x}$ are like terms.

Example 2. $\sqrt{3} - \frac{1}{\sqrt{3}} = \sqrt{3}\frac{\sqrt{3}}{\sqrt{3}} - \frac{1}{\sqrt{3}} = \frac{3}{\sqrt{3}} - \frac{1}{\sqrt{3}} = \frac{2}{\sqrt{3}} = \frac{2}{\sqrt{3}}\frac{\sqrt{3}}{\sqrt{3}} = \frac{2\sqrt{3}}{3}$
Notes: We multiplied $\sqrt{3}$ by $\frac{\sqrt{3}}{\sqrt{3}}$ in order to make a common denominator and multiplied $\frac{2}{\sqrt{3}}$ by $\frac{\sqrt{3}}{\sqrt{3}}$ in order to rationalize the denominator of the answer.

Example 3. $(2 + \sqrt{3})(6 - \sqrt{3}) = 2(6) - 2\sqrt{3} + 6\sqrt{3} - \sqrt{3}\sqrt{3} = 12 + 4\sqrt{3} - 3 = 9 + 4\sqrt{3}$.
Note: $-2\sqrt{3} + 6\sqrt{3} = 4\sqrt{3}$ because these are like terms (factor out the $\sqrt{3}$).

Example 4. $\sqrt{6x^7}\sqrt{15x^3} = \sqrt{(6)(15)x^7x^3} = \sqrt{90x^{10}} = x^5\sqrt{(9)(10)} = 3x^5\sqrt{10}$

Example 5. $\frac{\sqrt{6x^9}}{\sqrt{3x^3}} = \sqrt{\frac{6x^9}{3x^3}} = \sqrt{2x^6} = x^3\sqrt{2}$

2.8 Problems

Directions: Simplify each expression and express each answer in standard form.

(1) $9 + 2\sqrt{3} - 4 + 7\sqrt{3}$

(2) $6\sqrt{5} + 5\sqrt{6} - 2\sqrt{5} + 9\sqrt{6}$

(3) $x\sqrt{2} + 4\sqrt{x} + 8\sqrt{x} - 3x\sqrt{2}$

(4) $x^{3/2} + x - x\sqrt{x}$

(5) $5\sqrt{2} + \frac{6}{\sqrt{2}}$

(6) $\frac{7\sqrt{x}}{x} - \frac{4}{\sqrt{x}}$

(7) $\left(9 + \sqrt{5}\right)\left(3 - \sqrt{5}\right)$

(8) $\left(\sqrt{3} - \sqrt{2}\right)\left(\sqrt{3} - \sqrt{2}\right)$

(9) $\left(x + \sqrt{3}\right)\left(x - \sqrt{6}\right)$

(10) $\left(x + \sqrt{x}\right)\left(x + \sqrt{x}\right)$

(11) $\sqrt{x}\left(x\sqrt{x} - 4x + 3\sqrt{x} - 8\right)$

(12) $\left(x - 2\sqrt{x} + \sqrt{3}\right)\left(\sqrt{x} - \sqrt{3}\right)$

(13) $\left(x - \sqrt{x}\right)^3$

(14) $\left(3x + \sqrt{2}\right)^3$

(15) $\sqrt{2x^9}\sqrt{20x^5}$

(16) $\sqrt{6x^7}\sqrt{10x^5}\sqrt{15x^3}$

(17) $\frac{\sqrt{21x^7}}{\sqrt{7x}}$

(18) $\frac{\sqrt{30x^{21}}}{\sqrt{6x^{13}}}$

2.9 Equations with Exponents

To solve the simple equation $x^2 = 5$, square root both sides: $\sqrt{x^2} = \pm\sqrt{5}$. Simplify: $x = \pm\sqrt{5}$. The $\pm$ indicates that $x = -\sqrt{5}$ and $x = \sqrt{5}$ are both valid solutions. Observe that $\left(-\sqrt{5}\right)^2 = 5$ and $\left(\sqrt{5}\right)^2 = 5$. For a higher power, like $x^5 = 2$, take the corresponding root of both sides: $\sqrt[5]{x^5} = \sqrt[5]{2}$. Simplify: $x = \sqrt[5]{2}$. Check the answer by plugging it into the original equation: $\left(\sqrt[5]{2}\right)^5 = 2$. If the exponent is even, there are two solutions, indicated by a $\pm$ sign. If the exponent is odd, there is one real solution, which has the same sign as what is under the radical. Compare: $\sqrt[3]{27} = 3$, $\sqrt[3]{-27} = -3$, and $\sqrt[4]{81} = \pm 3$.

Not all equations are as simple as those discussed above. Many equations can be solved by isolating the unknown by combining like terms. For example, $3x^2 - 5 = 7 + x^2$ can be solved by adding 5 to both sides and subtracting x^2 from both sides to get $2x^2 = 12$, dividing by 2 on both sides to get $x^2 = 6$, and taking the square root of both sides: $x = \pm\sqrt{6}$. Check the answers with the original equation: $3\left(\pm\sqrt{6}\right)^2 - 5 = 3(6) - 5 = 18 - 5 = 13$ agrees with $7 + \left(\pm\sqrt{6}\right)^2 = 7 + 6 = 13$.

Example 1. $x^2 + 8 = 20$. Subtract 8 from both sides: $x^2 = 12$.
Square root both sides: $x = \pm\sqrt{12}$. Factor: $x = \pm\sqrt{4(3)} = \pm 2\sqrt{3}$.
Check: $\left(\pm 2\sqrt{3}\right)^2 + 8 = 2^2(3) + 8 = 4(3) + 8 = 12 + 8 = 20$.

Example 2. $-\frac{5}{x^3} = \frac{1}{2}$. Cross multiply: $-5(2) = x^3$. Simplify: $-10 = x^3$.
Cube root both sides: $-\sqrt[3]{10} = x$. Check: $\dfrac{5}{\left(-\sqrt[3]{10}\right)^3} = \dfrac{5}{-10} = -\dfrac{1}{2}$.

Example 3. $21 - x^4 = 2x^4$. Add x^4 to both sides: $21 = 3x^4$.
Divide by 3 on both sides: $7 = x^4$. Take the fourth root: $\pm\sqrt[4]{7} = x$.
Check: $21 - \left(\pm\sqrt[4]{7}\right)^4 = 21 - 7 = 14$ agrees with $2\left(\pm\sqrt[4]{7}\right)^4 = 2(7) = 14$.

2.9 Problems

Directions: Solve for the variable in each equation.

(1) $6 + 5x^2 = 41$

(2) $7x^2 - 24 = 4x^2 + 30$

(3) $10 - x^3 = x^3$

(4) $2x^5 - 59 = 6x^5 - 15$

(5) $9x^4 + 14 = 6x^4 + 62$

(6) $140 - x^2 = 20 + 5x^2$

(7) $\dfrac{8}{x^3} = \dfrac{2}{3}$

(8) $\dfrac{2}{x} = \dfrac{x}{3}$

(9) $\dfrac{1}{x^2} + \dfrac{1}{4} = \dfrac{1}{3}$

(10) $\dfrac{1}{6} - \dfrac{1}{x^3} = \dfrac{1}{2}$

2.10 Equations with Square Roots

If a variable is inside of a square root and the radical can be isolated (by combining like terms to place the square root by itself on one side of the equation), after isolating the radical, square both sides of the equation to remove the radical. For example, in the equation $5 + \sqrt{3x} = 17$, first subtract 5 from both sides to isolate the radical: $\sqrt{3x} = 12$. Then square both sides to remove the radical: $(\sqrt{3x})^2 = 12^2$. This simplifies to $3x = 144$. Finally, divide both sides by 3 to isolate the variable: $x = \frac{144}{3} = 48$. Check the answer by plugging it into the original equation: $5 + \sqrt{3(48)} = 5 + \sqrt{144} = 17$.

If the variable appears in two different radicals, first see if you can combine like terms. For example, the two radicals can be combined in $2\sqrt{x} = 6 - \sqrt{x}$ because $2\sqrt{x}$ and $-\sqrt{x}$ are like terms. Add $\sqrt{x}$ to both sides to get $3\sqrt{x} = 6$. Now divide both sides by 3 to get $\sqrt{x} = 2$ and square both sides to get $x = 2^2 = 4$. Check the answer by plugging it into the original equation: $2\sqrt{4} = 2(2) = 4$ agrees with $6 - \sqrt{4} = 6 - 2 = 4$.

If the variable appears in two different radicals and the terms can't be combined, you may be able to remove the radical by squaring both sides of the equation twice, but you may need to rearrange the terms after squaring the equation the first time. As an example, consider the equation $\sqrt{x + 19} + \sqrt{x} = 19$. First subtract $\sqrt{x}$ from both sides to get $\sqrt{x + 19} = 19 - \sqrt{x}$. Now square both sides. Apply the FOIL method to the right side: $x + 19 = 361 - 38\sqrt{x} + x$. Now subtract 361 and subtract x from both sides to isolate the radical term: $-342 = -38\sqrt{x}$. Divide by -38 on both sides: $9 = \sqrt{x}$. Square both sides: $81 = x$. Check the answer by plugging it into the original equation: $\sqrt{81 + 19} + \sqrt{81} = \sqrt{100} + 9 = 10 + 9 = 19$.

Example 1. $5 + \sqrt{11 - x} = 12$. Subtract 5 from both sides: $\sqrt{11 - x} = 7$. Square both sides: $11 - x = 49$. Subtract 11 from both sides: $-x = 38$. Multiply by -1 on both sides: $x = -38$.
Check: $5 + \sqrt{11 - (-38)} = 5 + \sqrt{49} = 5 + 7 = 12$.

Example 2. $2\sqrt{3x} = 12 - \sqrt{3x}$. Add $\sqrt{3x}$ to both sides: $3\sqrt{3x} = 12$.

Divide by 3 on both sides: $\sqrt{3x} = 4$. Square both sides: $3x = 16$.

Divide by 3 on both sides: $x = \frac{16}{3}$.

Check: $2\sqrt{3\left(\frac{16}{3}\right)} = 2\sqrt{16} = 2(4) = 8$ agrees with $12 - \sqrt{3\left(\frac{16}{3}\right)} = 12 - \sqrt{16} = 12 - 4 = 8$.

Example 3. $\sqrt{x+16} = 1 + \sqrt{x-7}$. Square both sides: $x + 16 = 1 + 2\sqrt{x-7} + x - 7$.

Simplify: $22 = 2\sqrt{x-7}$. Divide by 2 on both sides: $11 = \sqrt{x-7}$.

Square both sides: $121 = x - 7$. Add 7 to both sides: $128 = x$.

Check: $\sqrt{128+16} = \sqrt{144} = 12$ agrees with $1 + \sqrt{128-7} = 1 + \sqrt{121} = 1 + 11 = 12$.

2.10 Problems

Directions: Solve for the variable in each equation.

(1) $\sqrt{6x} - 3 = 9$

(2) $10 - \sqrt{x} = \sqrt{x}$

(3) $14 + \sqrt{x-3} = 20$

(4) $7\sqrt{2x} + 5 = 37 + 3\sqrt{2x}$

(5) $\sqrt{2x+7} - 4 = 7$

(6) $15 = 21 - \sqrt{45 - x^2}$

(7) $\dfrac{6}{\sqrt{x}} = \dfrac{12}{5}$

(8) $\dfrac{1}{\sqrt{2x}} - \dfrac{1}{18} = \dfrac{1}{9}$

(9) $4 - \dfrac{2}{\sqrt{x}} = \dfrac{3}{\sqrt{x}}$

(10) $\sqrt{3} - \sqrt{x} = \dfrac{1}{\sqrt{3}}$

(11) $\sqrt{x + 45} = 2 + \sqrt{x - 23}$

(12) $\sqrt{3x + 81} - 3 = \sqrt{3x}$

(13) $\sqrt{3} = \sqrt{x} - \sqrt{2}$

(14) $10 - \sqrt{2x + 11} = \sqrt{2x + 31}$

2.11 Equations with Fractional Exponents

If a variable with a fractional exponent is isolated, like $x^{2/3} = 9$, raise both sides of the equation to the reciprocal of the exponent. For example, raise both sides of $x^{2/3} = 9$ to the power of $\frac{3}{2}$ to get $\left(x^{2/3}\right)^{3/2} = 9^{3/2}$, which simplifies to $x = \pm 27$. Recall from Sec. 2.6 that $9^{3/2} = \left(9^{1/2}\right)^3 = \left(\pm\sqrt{9}\right)^3 = (\pm 3)^3 = \pm 27$. An **even** root like $9^{1/2}$ or like $16^{1/4}$ (where the denominator of the exponent is even) has both $\pm$ answers, whereas an **odd** root like $8^{1/3}$ or like $243^{1/5}$ (where the denominator of the exponent is odd) has one real answer which has the same sign as the number being rooted. Compare $8^{1/3} = 2$ with $(-8)^{1/3} = -2$ and with $4^{1/2} = \pm 2$. However, if the numerator is even, there will only be a positive real answer. For example, $(-8)^{2/3} = \left[(-8)^{1/3}\right]^2 = (-2)^2 = 4$.

Example 1. $x^{3/2} - 50 = 75$. Add 50 to both sides: $x^{3/2} = 125$.

Raise both sides to the power of two-thirds: $x = 125^{2/3} = \left(\sqrt[3]{125}\right)^2 = 5^2 = 25$.

Check: $25^{3/2} = \left(25^{1/2}\right)^3 - 50 = \left(\sqrt{25}\right)^3 - 50 = 5^3 - 50 = 125 - 50 = 75$.

Example 2. $48 + 2x^{4/3} = 5x^{4/3}$. Subtract $2x^{4/3}$ from both sides: $48 = 3x^{4/3}$.
Divide by 3 on both sides: $16 = x^{4/3}$. Raise both sides to the power of three-fourths:
$x = 16^{3/4} = \left(\pm 16^{1/4}\right)^3 = \left(\pm\sqrt[4]{16}\right)^3 = (\pm 2)^3 = \pm 8$.
Check: $48 + 2x^{4/3} = 48 + 2(\pm 8)^{4/3} = 48 + 2\left[(\pm 8)^{1/3}\right]^4 = 48 + 2\left(\sqrt[3]{\pm 8}\right)^4$
$= 48 + 2(\pm 2)^4 = 48 + 2(16) = 48 + 32 = 80$ agrees with $5x^{4/3} = 5(\pm 8)^{4/3}$
$= 5\left[(\pm 8)^{1/3}\right]^4 = 5\left(\sqrt[3]{\pm 8}\right)^4 = 5(\pm 2)^4 = 5(16) = 80$.

Example 3. $-\frac{4}{x^{3/5}} = \frac{1}{2}$. Cross multiply: $-8 = x^{3/5}$. Raise both sides to the power of five-thirds: $x = (-8)^{5/3} = \left[(-8)^{1/3}\right]^5 = \left(\sqrt[3]{-8}\right)^5 = (-2)^5 = -32$.
Check: $-\frac{4}{x^{3/5}} = -\frac{4}{(-32)^{3/5}} = -\frac{4}{\left[(-32)^{1/5}\right]^3} = -\frac{4}{\left(\sqrt[5]{-32}\right)^3} = -\frac{4}{(-2)^3} = -\frac{4}{-8} = \frac{1}{2}$.

2.11 Problems

Directions: Solve for the variable in each equation.

(1) $16 + x^{3/4} = 80$

(2) $8x^{2/5} - 12 = 5x^{2/5}$

(3) $x^{5/3} + 500 = 14 - x^{5/3}$

(4) $3x^{3/2} - 35 = 100 - 2x^{3/2}$

(5) $6x^{2/3} - 48 = 4x^{2/3} + 50$

(6) $x\sqrt{x} - 56 = 160$

(7) $\dfrac{54}{x^{4/3}} = \dfrac{2}{3}$

(8) $\dfrac{8}{x^{1/2}} = \dfrac{x^{1/3}}{4}$

(9) $\dfrac{1}{x^{2/3}} + \dfrac{1}{18} = \dfrac{1}{12}$

(10) $\dfrac{5}{12} - \dfrac{16}{x^{3/2}} = \dfrac{1}{6}$

3 COMPLEX NUMBERS

3.1 Imaginary Numbers

The square of a **real** number is always nonnegative. For example, $3^2 = 9$, $0^2 = 0$, and $(-1)^2 = 1$. The equation $x^2 = 1$ has two real solutions: $x = \pm 1$ because $1^2 = 1$ and $(-1)^2 = 1$. Similarly, the equation $x^2 = 0.25$ has two real solutions: $x = \pm 0.5$.

In contrast, the equation $x^2 = -1$ does NOT have a real solution. The square of a real number can't be negative. However, the equation $x^2 = -1$ does have an **imaginary** solution. An **imaginary** number is a number with a square that is negative. We use the symbol i to represent the imaginary number that solves the equation $x^2 = -1$. That is, $i^2 = -1$, such that $x = i$ solves the equation $x^2 = -1$. If you square root both sides of $i^2 = -1$, you get $i = \sqrt{-1}$.

Other imaginary numbers can be expressed in terms of i by factoring. For example, $\sqrt{-4} = \sqrt{4(-1)} = \sqrt{4}\sqrt{-1} = 2i$ and $\sqrt{-3} = \sqrt{3(-1)} = \sqrt{3}\sqrt{-1} = i\sqrt{3}$. It may help to review Sec. 2.2. For example, $\sqrt{-45} = \sqrt{9(5)(-1)} = \sqrt{9}\sqrt{5}\sqrt{-1} = 3i\sqrt{5}$.

Example 1. $\sqrt{-121} = \sqrt{121(-1)} = \sqrt{121}\sqrt{-1} = 11i$

Example 2. $\sqrt{-7} = \sqrt{7(-1)} = \sqrt{7}\sqrt{-1} = i\sqrt{7}$

Example 3. $\sqrt{-200} = \sqrt{100(2)(-1)} = \sqrt{100}\sqrt{2}\sqrt{-1} = 10i\sqrt{2}$

3.1 Problems

Directions: Express each answer in terms of i. Factor out any perfect squares.

(1) $\sqrt{-25}$ (2) $\sqrt{-17}$

(3) $\sqrt{-49}$

(4) $\sqrt{-23}$

(5) $\sqrt{-8}$

(6) $\sqrt{-81}$

(7) $\sqrt{-15}$

(8) $\sqrt{-48}$

(9) $\sqrt{-144}$

(10) $\sqrt{-63}$

(11) $\sqrt{-6}$

(12) $\sqrt{-75}$

(13) $\sqrt{-64}$

(14) $\sqrt{-42}$

(15) $\sqrt{-24}$

(16) $\sqrt{-256}$

(17) $\sqrt{-72}$

(18) $\sqrt{-245}$

(19) $\sqrt{-53}$

(20) $\sqrt{-361}$

3.2 Complex Numbers

A **complex** number has both real and imaginary parts. A complex number has the form $z = x + iy$, where x is the real part and y is the imaginary part. The notation $\text{Re}(z)$ means the **real** part of z, while the notation $\text{Im}(z)$ means the **imaginary** part of z. For example, for the complex number $z = -6 + 2i$, the real and imaginary parts are $\text{Re}(z) = -6$ and $\text{Im}(z) = 2$.

Example 1. For $z = 5 + 9i$, the real and imaginary parts are $\text{Re}(z) = 5$ and $\text{Im}(z) = 9$.

Example 2. For $w = 3 - 2i$, the real and imaginary parts are $\text{Re}(w) = 3$ and $\text{Im}(w) = -2$.

3.2 Problems

Directions: Find the real and imaginary parts of each complex number.

(1) Given $z = 4 + 5i$, find $\text{Re}(z)$ and $\text{Im}(z)$.

(2) Given $w = 2 + i$, find $\text{Re}(w)$ and $\text{Im}(w)$.

(3) Given $u = -7 + 6i$, find $\text{Re}(u)$ and $\text{Im}(u)$.

(4) Given $z = -3 - 3i$, find $\text{Re}(z)$ and $\text{Im}(z)$.

(5) Given $v = 8i$, find $\text{Re}(v)$ and $\text{Im}(v)$.

(6) Given $t = 5 - i$, find $\text{Re}(t)$ and $\text{Im}(t)$.

(7) Given $w = -10 + 2i$, find $\text{Re}(w)$ and $\text{Im}(w)$.

(8) Given $z = 2 - \frac{i}{2}$, find $\text{Re}(z)$ and $\text{Im}(z)$.

(9) Given $d = 12{,}073$, find $\text{Re}(d)$ and $\text{Im}(d)$.

(10) Given $q = 0.4 + i\sqrt{2}$, find $\text{Re}(q)$ and $\text{Im}(q)$.

3.3 Adding and Subtracting Complex Numbers

To add or subtract complex numbers, add or subtract their real and imaginary parts. For example, $(3 + 4i) + (8 + 2i) = (3 + 8) + (4 + 2)i = 11 + 6i$.

Example 1. $(8 + 3i) + (9 - 7i) = (8 + 9) + (3 - 7)i = 17 - 4i$

Example 2. $(5 + 2i) - (6 - 4i) = (5 - 6) + [2 - (-4)]i = -1 + (2 + 4)i = -1 + 6i$

3.3 Problems

Directions: Carry out arithmetic with the complex numbers.

(1) What is $5 + 9i$ plus $6 + 4i$?

(2) What is $6 - 2i$ plus $4 + 7i$?

(3) What is $14 + 7i$ plus $16 - 3i$?

(4) What is $11 - 3i$ plus $6 - 6i$?

(5) What is $15 - 8i$ plus $-2 + i$?

(6) What is $8 + 6i$ minus $9 + 3i$?

(7) What is $4 + 7i$ minus $10 - 2i$?

(8) What is $12 - 5i$ minus $3 + 4i$?

(9) What is $16 - 12i$ minus $9 - 7i$?

(10) What is $11 + 14i$ minus $-8 + 6i$?

3.4 Multiplying Complex Numbers

To multiply two complex numbers, apply the FOIL method. Note that $i^2 = -1$. For example, $(4 + 2i)(3 + 5i) = 4(3) + 4(5i) + 2i(3) + 2i(5i) = 12 + 20i + 6i + 10i^2 = 12 + 26i + 10(-1) = 12 + 26i - 10 = 2 + 26i$.

The powers of the imaginary number form a repeating pattern: $i^0 = 1, i^1 = i, i^2 = -1,$ $i^3 = -i, i^4 = 1, i^5 = i, i^6 = -1, i^7 = -i, i^8 = 1, i^9 = i, i^{10} = -1$, etc. In this pattern, i^n equals 1 whenever n is evenly divisible by 4.

Example 1. $(2 + 7i)(4 - i) = 2(4) + 2(-i) + 7i(4) + 7i(-i) = 8 - 2i + 28i - 7(-1) = 8 + 26i + 7 = 15 + 26i$

Example 2. $i^{27} = i^{24}i^3 = (i^4)^6 i^3 = (1)^6(-i) = 1(-i) = -i$
Alternative solution: Since $27 \div 4$ has a remainder of 3, $i^{27} = i^3 = -i$.

3.4 Problems

Directions: Carry out arithmetic with the complex numbers.

(1) $(6 + 3i)(8 + 4i)$

(2) $(3 + i)(9 - i)$

(3) $(5 - 6i)(3 + 2i)$

(4) $(8 - 5i)(9 - 8i)$

(5) $(-5 + 9i)(7 - 3i)$

(6) $-7i(4 - i)$

(7) $(1 - i)(-1 + i)$

(8) $(4 + 6i)(4 + 6i)$

(9) $(7 + 9i)(7 - 9i)$

(10) $(8 - 5i)^2$

(11) $(3 + 2i)^3$

(12) i^{46} (13) i^{101}

(14) i^{75} (15) i^{68}

3.5 Complex Conjugates

The **complex conjugate** of a complex number changes the sign of the imaginary part. An asterisk (*) is used to indicate a complex conjugate. For example, for the complex number $z = 5 + 2i$, the complex conjugate is $z^* = 5 - 2i$. The complex conjugate is important because of its role in the modulus squared (Sec. 3.6).

Example 1. Given $z = 6 + 4i$, the complex conjugate is $z^* = 6 - 4i$.

Example 2. Given $w = 7 - 2i$, the complex conjugate is $w^* = 7 + 2i$.

3.5 Problems

Directions: Find the complex conjugate of each complex number.

(1) $z = 9 + 7i$

(2) $w = 5 - 8i$

(3) $u = -1 + i$

(4) $z = -3 - 6i$

(5) $v = 16i$

(6) $t = 4 - i$

(7) $z = \sqrt{2} + i\sqrt{3}$

(8) $w = 3.25 + 1.75i$

(9) $p = 3 - \sqrt{5} + 6i - i\sqrt{5}$

(10) $q = -i$

(11) $d = 12$

(12) $z = \frac{1}{2} + \frac{i\sqrt{3}}{2}$

3.6 The Modulus

The **modulus squared** is formed by multiplying a complex number and its complex conjugate together. The notation $|z|^2$ refers to the modulus squared.

$$|z|^2 = zz^* = z^*z$$

For a complex number of the form $z = x + iy$, the complex conjugate is:

$$|z|^2 = zz^* = (x + iy)(x - iy) = x^2 - ixy + ixy - y^2(-1) = x^2 + y^2$$

The modulus squared equals the sum of the squares of the real and imaginary parts. Note that the modulus squared is always a real number, since no i appears in $x^2 + y^2$. The modulus squared is also nonnegative, since x^2 and y^2 are each nonnegative.

The **modulus** of a complex number refers to the positive root of the modulus squared:

$$|z| = \sqrt{|z|^2} = \sqrt{zz^*} = \sqrt{x^2 + y^2}$$

Example 1. Given $z = 4 + 3i$, the modulus is $|z| = \sqrt{4^2 + 3^2} = \sqrt{16 + 9} = \sqrt{25} = 5$.

Example 2. Given $w = 2 - 4i$, the modulus is $|w| = \sqrt{2^2 + (-4)^2} = \sqrt{4 + 16} = \sqrt{20} = \sqrt{4(5)} = \sqrt{4}\sqrt{5} = 2\sqrt{5}$.

3.6 Problems

Directions: Find the modulus of each complex number.

(1) $z = 8 + 6i$

(2) $w = \sqrt{3} - i$

(3) $u = -3 + 6i$

(4) $z = -\sqrt{2} - i\sqrt{2}$

(5) $v = 12 + 5i$

(6) $t = 7i$

(7) $z = \sqrt{6} - i\sqrt{3}$

(8) $w = -\frac{1}{2} + \frac{i}{4}$

(9) $p = -9 - 3i$

(10) $q = 8 - 15i$

(11) $d = \sqrt{11} + i$

(12) $z = -5 + 10i$

(13) $w = 12 - i\sqrt{6}$

(14) $c = 1 - i$

(15) $z = -3\sqrt{2} - 4i\sqrt{5}$

3.7 Complex Fractions

If a complex fraction has a complex denominator, express the fraction in standard form as follows. Multiply the numerator and denominator by the complex conjugate of the denominator in order to make the denominator real. (This is similar in spirit to Sec. 2.5, except that with complex numbers you multiply by the complex conjugate.) For example, multiply the numerator and denominator of $\frac{4-3i}{5+2i}$ each by $5 - 2i$:

$$\frac{4-3i}{5+2i} = \frac{4-3i}{5+2i}\left(\frac{5-2i}{5-2i}\right) = \frac{4(5)+4(-2i)-3i(5)-3i(-2i)}{5^2+2^2}$$

$$= \frac{20-8i-15i+6(-1)}{25+4} = \frac{20-23i-6}{29} = \frac{14-23i}{29}$$

In the denominator, $(5+2i)(5-2i)$ forms the modulus squared. We applied the rule $(x+iy)(x-iy) = x^2 + y^2$ from Sec. 3.6 to the denominator. In the numerator, note that $-3i(-2i) = 6i^2 = 6(-1) = -6$.

The rule for expressing a complex fraction in standard form is also the rule for how to divide two complex numbers. Observe that $(4-3i) \div (5+2i) = \frac{4-3i}{5+2i}$. Any division problem can similarly be expressed as a fraction.

Example 1. $\frac{4+5i}{3-2i} = \frac{4+5i}{3-2i}\left(\frac{3+2i}{3+2i}\right) = \frac{4(3)+4(2i)+5i(3)+5i(2i)}{3^2+2^2} = \frac{12+8i+15i+10(-1)}{9+4} = \frac{12+23i-10}{13} = \frac{2+23i}{13}$

Example 2. $\frac{7+4i}{3+6i} = \frac{7+4i}{3+6i}\left(\frac{3-6i}{3-6i}\right) = \frac{7(3)+7(-6i)+4i(3)+4i(-6i)}{3^2+6^2} = \frac{21-42i+12i-24(-1)}{9+36} = \frac{21-30i+24}{45} =$

$\frac{45-30i}{45} = \frac{3-2i}{3}$ We divided the numerator and denominator by 15 to reduce the fraction.

3.7 Problems

Directions: Express each complex fraction in standard form.

(1)

$$\frac{2+4i}{5+3i}$$

(2)

$$\frac{7 + 3i}{2 - 6i}$$

(3)

$$\frac{9 - 4i}{5 + 8i}$$

(4)

$$\frac{10 - 15i}{5 - 2i}$$

(5)

$$\frac{1 + i}{1 - i}$$

(6)

$$\frac{6}{7 + 9i}$$

(7)

$$\frac{6 + 3i}{4 - 8i}$$

(8)

$$\frac{2 - i}{3 - 7i}$$

(9)

$$\frac{8}{9 - 3i}$$

(10)

$$\frac{6 - 5i}{8 + 9i}$$

3.8 The Complex Plane

A complex number $z = x + iy$ can be graphed in the **complex plane** (also called the **Argand plane**). The real part and imaginary part form an ordered pair (x, y).

- The x-axis is the real axis. A purely real number lies on the x-axis.
- The y-axis is the imaginary axis. A purely imaginary number lies on the y-axis.
- A complex number with both real and imaginary parts lies between the axes.

The modulus (Sec. 3.6) provides the distance from the origin to (x, y).

$$|z| = \sqrt{|z|^2} = \sqrt{zz^*} = \sqrt{x^2 + y^2}$$

Example 1. Graph the point $4 - 3i$ in the complex plane and determine the distance from the origin to this point.

The real part is 4 and the imaginary part is -3. Plot the point $(4, -3)$. This point is 4 units to the right and 3 units down.

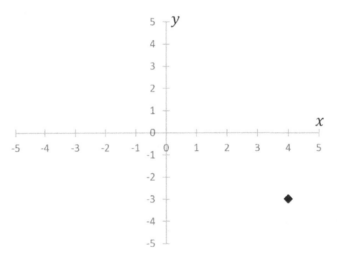

The distance from the origin to $(4, -3)$ is equal to the modulus of the complex number. This can be visualized as the hypotenuse of a right triangle with a horizontal side of 4 (the real part) and a vertical side of -3 (the imaginary part).

$$|z| = \sqrt{x^2 + y^2} = \sqrt{4^2 + (-3)^2} = \sqrt{16 + 9} = \sqrt{25} = 5$$

3.8 Problems

Directions: Plot each complex number in the complex plane and determine its distance from the origin.

(1) $z = 9 + 7i$

(2) $w = 8 - 6i$

(3) $u = -3\sqrt{2} + 3i\sqrt{6}$

(4) $v = 8i$

(5) $t = -6 - 3i$

(6) $c = 2\sqrt{2} + 2i\sqrt{2}$

(7) $d = -3$

(8) $q = 4 - 8i$

3.9 Complex Functions

If we allow the range of a function to include complex numbers, the domain can be expanded beyond what we had considered in Chapter 1. For example, the domain of $f(x) = \sqrt{x}$ can be expanded to include negative values, provided that the range allows for complex numbers. When $x = -9$, for example, $f(-9) = \sqrt{-9} = \sqrt{9}\sqrt{-1} = 3i$ is an imaginary value.

The domain of a function can similarly be expanded to include complex numbers. For example, if the domain of $f(x) = x^2$ includes all complex numbers, then the range can include complex numbers like $f(2 + 3i) = (2 + 3i)^2 = 2^2 + 12i + 9(-1) = -5 + 12i$.

Example 1. Given $f(x) = \sqrt{3 - x}$, $f(7) = \sqrt{3 - 7} = \sqrt{-4} = \sqrt{4}\sqrt{-1} = 2i$.

Example 2. Given $f(x) = 5x - 2$, $f(3 - 4i) = 5(3 - 4i) - 2 = 15 - 20i - 2 = 13 - 20i$.

3.9 Problems

Directions: Evaluate each function at the specified value.

(1) Given $f(x) = \sqrt{11 - x^2}$, find $f(6)$.   (2) Given $g(y) = 4y + 9$, find $g(7 - 5i)$.

(3) Given $y(x) = 8x^2 - 5ix$, find $y(2i)$.   (4) Given $p(q) = 2q^2$, find $p(9 - 6i)$.

(5) Given $h(t) = \frac{2t - 5i}{4t + 3i}$, find $h(5 + 2i)$.   (6) Given $f(x) = x^{3/2}$, find $f(-16)$.

3.10 Complex Solutions

An equation like $x^2 = -4$ doesn't have a <u>real</u> solution because the square of a real number can't be negative, but it has a <u>complex</u> solution: $x = \pm\sqrt{-4} = \pm 2i$. Check the answer by plugging it into the original equation: $(\pm 2i)^2 = 2^2 i^2 = 4(-1) = -4$.

Example 1. $2x^2 + 17 = -19$. Subtract 17 from both sides: $2x^2 = -36$.
Divide by 2 on both sides: $x^2 = -18$. Square root both sides: $x = \pm\sqrt{-18}$.
Factor out the perfect square: $x = \pm\sqrt{9}\sqrt{-2} = \pm 3i\sqrt{2}$.
Check: $2\left(\pm 3i\sqrt{2}\right)^2 + 17 = 2(3)^2 i^2\left(\sqrt{2}\right)^2 + 17 = 2(9)(-1)(2) + 17 = -36 + 17 = -19$.

Example 2. $3z - 6i = 5z - 12$. Add 12 and subtract $3z$ from both sides: $12 - 6i = 2z$.
Divide by 2 on both sides: $6 - 3i = z$.
Check: $3(6 - 3i) - 6i = 18 - 9i - 6i = 18 - 15i$ agrees with $5(6 - 3i) - 12 = 30 - 15i - 12 = 18 - 15i$.

3.10 Problems

Directions: Solve for the variable in each equation.

(1) $76 - 5x^2 = 256$

(2) $9z + 14 = 7z - 18i$

(3) $11i - 19 - 2z = 13 + 35i - 6z$

(4) $24 - x^2 = x^2 + 120$

(5) $350 - 4x^2 = 56 - 7x^2$

(6) $\frac{2+i}{z} = \frac{2-i}{3-i}$

(7) $12 - 3(6i - z) = 6i - z$

(8) $3x^2 + 300 = 57$

(9) $6z^* - 12i = 2z^* + 8$

(10) $\frac{18}{x} = -\frac{x}{4}$

(11) $\frac{x^3}{8} = -32x$

(12) $2x^2 + 350 = 150$

(13) $\frac{1}{2i} - \frac{1}{z} = \frac{1}{3}$

(14) $\frac{3}{z+3i} = \frac{7}{z-5i}$

4 QUADRATICS

4.1 The Quadratic Formula

A **quadratic** expression is a polynomial with degree two. It has the form $ax^2 + bx + c$. A **quadratic** equation includes a quadratic expression and an equal sign. The **standard form** for a quadratic equation is $ax^2 + bx + c = 0$. If a quadratic equation is not yet in standard form, put it in standard form by combining like terms and rearranging the terms so that the squared term is first, the linear term is second, and the constant term is third on the same side of the equation. It is okay to write the equation in the form $0 = ax^2 + bx + c$. Once the quadratic equation is in standard form, one way to solve the quadratic equation is to use the **quadratic formula**:

$$x = \frac{-b \pm \sqrt{b^2 - 4ac}}{2a}$$

You can verify that the quadratic formula correctly solves the quadratic equation by replacing x with $\frac{-b \pm \sqrt{b^2 - 4ac}}{2a}$ in the quadratic equation and simplifying. If you do this correctly, everything will cancel out such that $ax^2 + bx + c$ equals zero.

Example 1. $2x^2 - 7x - 30 = 0$. This quadratic equation is already in standard form. Identify the coefficients: $a = 2$, $b = -7$, and $c = -30$.

$$x = \frac{-b \pm \sqrt{b^2 - 4ac}}{2a} = \frac{-(-7) \pm \sqrt{(-7)^2 - 4(2)(-30)}}{2(2)}$$

$$x = \frac{7 \pm \sqrt{49 + 240}}{4} = \frac{7 \pm \sqrt{289}}{4} = \frac{7 \pm 17}{4}$$

$$x = \frac{7 + 17}{4} = \frac{24}{4} = 6 \quad \text{or} \quad x = \frac{7 - 17}{4} = \frac{-10}{4} = -\frac{5}{2}$$

Check: $2(6)^2 - 7(6) - 30 = 2(36) - 42 - 30 = 72 - 72 = 0$ and $2\left(-\frac{5}{2}\right)^2 - 7\left(-\frac{5}{2}\right) - 30 = 2\left(\frac{25}{4}\right) + \frac{35}{2} - 30 = \frac{25}{2} + \frac{35}{2} - 30 = \frac{60}{2} - 30 = 30 - 30 = 0$.

Example 2. $4x - 3x^2 = 10x + 2$. Subtract $4x$ from both sides: $-3x^2 = 6x + 2$. Add $3x^2$ to both sides: $0 = 3x^2 + 6x + 2$. This is equivalent to $3x^2 + 6x + 2 = 0$. Identify the coefficients: $a = 3$, $b = 6$, and $c = 2$.

$$x = \frac{-b \pm \sqrt{b^2 - 4ac}}{2a} = \frac{-6 \pm \sqrt{6^2 - 4(3)(2)}}{2(3)} = \frac{-6 \pm \sqrt{36 - 24}}{6}$$

$$x = \frac{-6 \pm \sqrt{12}}{6} = \frac{-6 \pm \sqrt{4}\sqrt{3}}{6} = \frac{-6 \pm 2\sqrt{3}}{6} = \frac{-3 \pm \sqrt{3}}{3}$$

$$x = \frac{-3 + \sqrt{3}}{3} \approx -0.4226 \quad \text{or} \quad x = \frac{-3 - \sqrt{3}}{3} \approx -1.577$$

Calculator check: $4(-0.4226) - 3(-0.4226)^2 \approx -2.226$ agrees with $10(-0.4226) + 2 \approx -2.226$ and $4(-1.577) - 3(-1.577)^2 \approx -13.77$ agrees with $10(-1.577) + 2 \approx -13.77$.

4.1 Problems

Directions: Use the quadratic formula to solve each quadratic equation.

(1) $x^2 - 13x + 40 = 0$

(2) $3x^2 + 25x = 18$

(3) $2x - 6 + 3x^2 = 8x$

(4) $30x + 4x^2 = 7x - 28$

(5) $33x - 9 = 18x^2 - 4$

(6) $12x + 2x^2 = -8$

(7) $5x^2 + 24x = 4x - 10$

(8) $9x^2 - 45x = 5x^2 + 36$

(9) $4x^2 + 3 = 3x(x + 2)$

(10) $11x + 18 = 40x^2 + 40x$

(11) $15x^2 + 31x = 6x^2 - 12$

(12) $(2x + 8)(x - 6) = -44$

4.2 Factoring Simple Quadratics

The problems in this section are simple because the coefficient of x^2 equals one. (In Sec. 4.3, we'll factor quadratics where the coefficient of x^2 isn't one.) For a quadratic expression that has the simple form $x^2 + bx + c$ (where $a = 1$), the expression can be factored by finding two numbers that have a **sum** of b and a **product** of c. For example, $b = 10$ and $c = 24$ in $x^2 + 10x + 24$. Which two numbers have a sum of 10 and a product of 24? Start by thinking about which numbers multiplied together make 24. The pairs are 24×1, 12×2, 8×3, and 6×4. For which pair do the numbers add up to 10? The answer is 6 and 4 because they have a sum of $6 + 4 = 10$ and a product of $6 \times 4 = 24$. Therefore, the expression $x^2 + 10x + 24$ factors as $(x + 4)(x + 6)$. Check the answer using the FOIL method:

$$(x + 4)(x + 6) = x^2 + 6x + 4x + 24 = x^2 + 10x + 24$$

If c is negative, one of the factors of c will be negative. For example, for $c = -8$, the possible pairs are -8×1, -4×2, -2×4, and -1×8. Note that -8×1 is different from -1×8 because -8 and 1 have a sum of -7, whereas -1 and 8 have a sum of 7.

If c is positive, but b is negative, then both factors will be negative. For example, for $c = 6$ and $b = -5$, consider the factors $-6 \times (-1)$ and $-3 \times (-2)$.

Example 1. $x^2 + 10x + 16$. The factors of $c = 16$ are 16×1, 8×2, and 4×4. The pair 8 and 2 add up to $b = 10$. The answer is $(x + 8)(x + 2)$.
Check: $(x + 8)(x + 2) = x^2 + 2x + 8x + 16 = x^2 + 10x + 16$.

Example 2. $x^2 - 2x - 15$. The factors of $c = -15$ are -15×1, -5×3, -3×5, and -1×15. The pair -5 and 3 add up to $b = -2$. The answer is $(x - 5)(x + 3)$.
Check: $(x - 5)(x + 3) = x^2 + 3x - 5x - 15 = x^2 - 2x - 15$.

Example 3. $x^2 - 5x + 6$. Since $b = -5 < 0$ and $c = 6 > 0$, the factors of $c = 6$ are $-6 \times (-1)$ and $-3 \times (-2)$. The pair -3 and -2 add up to $b = -5$. The answer is $(x - 3)(x - 2)$.
Check: $(x - 3)(x - 2) = x^2 - 2x - 3x + 6 = x^2 - 5x + 6$.

4.2 Problems

Directions: Factor each quadratic expression.

(1) $x^2 + 8x - 9$

(2) $x^2 + 13x + 30$

(3) $x^2 - 8x + 16$

(4) $x^2 - 7x - 18$

(5) $x^2 + 9x - 36$

(6) $x^2 + 12x + 32$

(7) $x^2 - 36$

(8) $x^2 - 14x + 40$

(9) $x^2 - 7x - 60$

(10) $x^2 + 17x + 16$

(11) $x^2 - 16x + 28$

(12) $x^2 + 17x - 60$

4.3 Factoring Quadratics

Some quadratic equations can be solved more efficiently by factoring than by using the quadratic formula. When an equation of the form $ax^2 + bx + c = 0$ factors nicely (note that, unlike the simpler quadratics of Sec. 4.2, in this section $a \neq 1$), first think about the factors of a and c, and then determine which combination of factors for each of these will make the correct b. For example, for $10x^2 - 3x - 18$, the factors of $a = 10$ include 10×1 and 5×2 while the factors of c include -18×1, -9×2, -6×3, -3×6, -2×9, and -1×18. (Beware that the rules for the signs from Sec. 4.2 do not always apply when $a \neq 1$. You need to reason out the signs for each problem.) Which pair of factors for a and which pair of factors for c combine together to make $b = -3$? When we pair the factors together, we match each factor of a's pair with one factor of c's pair, then add these together. The answer involves the 5 and 2 for a and the -3 and 6 for c: $5 \times (-3) + 2 \times 6 = -15 + 12 = -3$. The expression $10x^2 - 3x - 18$ factors as $(2x - 3)(5x + 6)$. Check the answer using the FOIL method:
$$(2x - 3)(5x + 6) = 10x^2 + 12x - 15x - 18 \doteq 10x^2 - 3x - 18$$

When this is new to a student, it usually seems more difficult than it actually is. After enough practice, students who become fluent with factoring quadratics are usually pretty good at thinking of the factors of a and c, and develop a good intuition for how to combine them. When a student finds this method to be difficult, one way to try to work it is to try to FOIL out different expressions using factors for a and c. For example, for $10x^2 - 3x - 18$, you can try $(10x - 1)(x + 18)$, $(5x - 1)(2x + 18)$, etc. If you do this blindly, there can be several possibilities to work out (especially if a and c each have many possible factors). When you can use logic and reason to rule out some of the possibilities (based on the value of b), this helps to speed things up. What it really takes to develop fluency with factoring is practice (and attempting to think through the logic even if you are struggling).

Factoring can be used to solve a quadratic equation. For example, given the equation $10x^2 - 3x - 18 = 0$, we can factor this as $(2x - 3)(5x + 6) = 0$, which means that either $2x - 3 = 0$ or $5x + 6 = 0$. Now solve for x for each case. Either $x = \frac{3}{2}$ or $x = -\frac{6}{5}$.

To check the answers, plug them into the original equation: $10 \left(\frac{3}{2}\right)^2 - 3 \left(\frac{3}{2}\right) - 18 =$

$10 \left(\frac{9}{4}\right) - \frac{9}{2} - 18 = \frac{45}{2} - \frac{9}{2} - \frac{36}{2} = 0$ and $10 \left(-\frac{6}{5}\right)^2 - 3 \left(-\frac{6}{5}\right) - 18 = 10 \left(\frac{36}{25}\right) + \frac{18}{5} - 18$

$= \frac{360}{25} + \frac{18}{5} - 18 = \frac{72}{5} + \frac{18}{5} - \frac{90}{5} = 0.$

Example 1. $8x^2 + 30x + 18 = 0$. The factors of $a = 8$ are 8×1 and 4×2. The factors of $c = 18$ are $18 \times 1, 9 \times 2$, are 6×3. The pair 4 and 2 for a and the pair 6 and 3 for c make $b = 30$ since $4 \times 6 + 2 \times 3 = 30$. The given equation factors as $(4x + 3)(2x + 6)$ $= 0$. Either $4x + 3 = 0$ or $2x + 6 = 0$, which means $x = -\frac{3}{4}$ or $x = -3$.

Check: $8 \left(-\frac{3}{4}\right)^2 + 30 \left(-\frac{3}{4}\right) + 18 = 8 \left(\frac{9}{16}\right) - \frac{90}{4} + 18 = \frac{9}{2} - \frac{45}{2} + \frac{36}{2} = 0$ and

$8(-3)^2 + 30(-3) + 18 = 8(9) - 90 + 18 = 72 - 90 + 18 = 0.$

Example 2. $21x^2 - 47x + 20 = 0$. The factors of $a = 21$ are 21×1 and 7×3. The factors of $c = 20$ are $-20 \times (-1), -10 \times (-2)$, and $-5 \times (-4)$. The pair 7 and 3 for a and the pair -5 and -4 for c make $b = -47$ since $7 \times (-5) + 3 \times (-4) = -47$. The given equation factors as $(7x - 4)(3x - 5) = 0$. Either $7x - 4 = 0$ or $3x - 5 = 0$, which means $x = \frac{4}{7}$ or $x = \frac{5}{3}$.

Check: $21 \left(\frac{4}{7}\right)^2 - 47 \left(\frac{4}{7}\right) + 20 = 21 \left(\frac{16}{49}\right) - \frac{188}{7} + 20 = \frac{48}{7} - \frac{188}{7} + \frac{140}{7} = 0$ and

$21 \left(\frac{5}{3}\right)^2 - 47 \left(\frac{5}{3}\right) + 20 = 21 \left(\frac{25}{9}\right) - \frac{235}{3} + 20 = \frac{175}{3} - \frac{235}{3} + \frac{60}{3} = 0.$

4.3 Problems

Directions: Solve each quadratic equation by factoring.

(1) $6x^2 - 36x + 48 = 0$

(2) $15x^2 - 27x - 54 = 0$

(3) $8x^2 + 93x - 36 = 0$

(4) $9x^2 + 33x + 24 = 0$

(5) $12x^2 - 16x - 35 = 0$

(6) $9x^2 - 50x + 25 = 0$

(7) $48x^2 + 112x + 9 = 0$

(8) $60x^2 + 4x - 45 = 0$

(9) $9x^2 - 60x + 36 = 0$

(10) $10x^2 + 43x + 42 = 0$

(11) $6x^2 + 4x - 16 = 0$

(12) $10x^2 + 33x - 72 = 0$

4.4 Completing the Square

Compare $4x^2 + 20x + 10$ with $(2x + 5)(2x + 5) = 4x^2 + 20x + 25$. It would factor nicely if the 10 were a 25. What we can do is replace the 10 with $25 - 15$. This is called **completing the square**.

$$4x^2 + 20x + 10 = 4x^2 + 20x + 25 - 15 = (2x + 5)^2 - 15$$

In general, the expression $ax^2 + bx + c$ can be expressed as:

$$ax^2 + bx + c = \left(x\sqrt{a} + \frac{b\sqrt{a}}{2a}\right)^2 + c - \frac{b^2}{4a}$$

For example, $a = 4$, $b = 20$, and $c = 10$ for $4x^2 + 20x + 10$:

$$\left[x\sqrt{4} + \frac{20\sqrt{4}}{2(4)}\right]^2 + 10 - \frac{20^2}{4(4)} = \left[2x + \frac{20(2)}{8}\right]^2 + 10 - \frac{400}{16} = (2x + 5)^2 - 15$$

Example 1. $x^2 + 8x + 7$. Compare with $(x + 4)^2 = x^2 + 8x + 16$. Rewrite 7 as $16 - 9$.

$$x^2 + 8x + 7 = x^2 + 8x + 16 - 9 = (x + 4)^2 - 9$$

Check: $(x + 4)^2 - 9 = x^2 + 8x + 16 - 9 = x^2 + 8x + 7$.

Example 2. $9x^2 - 12x + 15$. Compare with $(3x - 2)^2 = 9x^2 - 12x + 4$. Rewrite 15 as $4 + 11$.

$$9x^2 - 12x + 15 = 9x^2 - 12x + 4 + 11 = (3x - 2)^2 + 11$$

Check: $(3x - 2)^2 + 11 = 9x^2 - 12x + 4 + 11 = 9x^2 - 12x + 15$.

4.4 Problems

Directions: Complete the square for each expression.

(1) $x^2 + 16x + 9$

(2) $16x^2 + 24x + 20$

(3) $x^2 - 12x + 18$

(4) $49x^2 + 126x + 90$

(5) $4x^2 - 20x - 6$

(6) $25x^2 - 40x + 50$

(7) $64x^2 + 48x - 4$

(8) $9x^2 - 30x + 42$

(9) $36x^2 + 84x + 48$

(10) $81x^2 - 36x - 4$

(11) $49x^2 - 112x + 76$

(12) $16x^2 + 72x - 18$

4.5 The Discriminant

The **discriminant** refers to the part of the quadratic formula that is inside of the radical: $b^2 - 4ac$. If a, b, and c are real numbers, the discriminant can be used to determine the nature of the solutions to a quadratic solution. If $b^2 - 4ac > 0$, there are **two** distinct **real** answers, if $b^2 - 4ac = 0$, there is **one** distinct **real** answer, and if $b^2 - 4ac < 0$, there are **two** distinct **complex** answers.

Example 1. $2x^2 + 5x - 3 = 0$. Since $b^2 - 4ac = 5^2 - 4(2)(-3) = 25 + 24 = 49 > 0$, this quadratic equation has **two** distinct **real** answers.

Example 2. $3x^2 - 12x + 12 = 0$. Since $b^2 - 4ac = (-12)^2 - 4(3)(12) = 144 - 144 = 0$, this quadratic equation has **one** distinct **real** answer (called a **double root**).

Example 3. $5x^2 + 8x + 4 = 0$. Since $b^2 - 4ac = 8^2 - 4(5)(4) = 64 - 80 = -16 < 0$, this quadratic equation has **two** distinct **complex** answers.

4.5 Problems

Directions: Use the discriminant to determine the nature of the answers.

(1) $2x^2 - 7x + 6 = 0$

(2) $x^2 + 5x + 7 = 0$

(3) $8x^2 + 12x - 5 = 0$

(4) $4x^2 - 20x + 25 = 0$

(5) $-3x^2 + 10x - 8 = 0$

(6) $x^2 - 2x - 1 = 0$

(7) $7x^2 + 14x + 7 = 0$

(8) $2x^2 - 9x + 11 = 0$

(9) $-8x^2 - 8x + 2 = 0$

(10) $-5x^2 - 6x - 2 = 0$

4.6 Complex Solutions to the Quadratic

The solutions to the quadratic equation are complex when the discriminant (Sec. 4.5) is negative. In this case, factor out $i = \sqrt{-1}$ (Sec. 3.1) to express the answer in terms of the imaginary number i.

Example 1. $x^2 - 6x + 25 = 0$. Identify $a = 1$, $b = -6$, and $c = 25$.

$$x = \frac{-b \pm \sqrt{b^2 - 4ac}}{2a} = \frac{-(-6) \pm \sqrt{(-6)^2 - 4(1)(25)}}{2(1)} = \frac{6 \pm \sqrt{36 - 100}}{2}$$

$$x = \frac{6 \pm \sqrt{-64}}{2} = \frac{6 \pm \sqrt{(64)(-1)}}{2} = \frac{6 \pm \sqrt{64}\sqrt{-1}}{2} = \frac{6 \pm 8i}{2} = 3 \pm 4i$$

The two answers, $x = 3 + 4i$ and $x = 3 - 4i$, are complex conjugates of one another. Check: $(3 \pm 4i)^2 - 6(3 \pm 4i) + 25 = 9 \pm 24i + 16(-1) - 18 \mp 24i + 25 = 9 - 16 - 18 + 25 = 34 - 34 = 0$. Note: $\mp 24i$ and $\pm 24i$ have opposite signs.

4.6 Problems

Directions: Use the quadratic formula to solve each quadratic equation.

(1) $x^2 - 2x + 2 = 0$

(2) $x^2 - 6x + 13 = 0$

(3) $x^2 + 4x + 29 = 0$

(4) $-4x^2 + 4x - 5 = 0$

(5) $2x^2 - 24x + 80 = 0$

(6) $4x^2 + 8x + 5 = 0$

(7) $-4x^2 - 12x - 73 = 0$

(8) $3x^2 - 30x + 102 = 0$

(9) $-9x^2 + 12 - 5 = 0$

(10) $25x^2 + 100x + 104 = 0$

(11) $2x^2 - 28x + 116 = 0$

(12) $-6x^2 - 30x - 159 = 0$

4.7 Cross Multiplying

An equation of the form $\frac{x+3}{6} = \frac{x}{x-2}$ can be solved by **cross multiplying**, which has the same result as multiplying both sides by the denominators: $(x+3)(x-2) = 6x$. Use the FOIL method: $x^2 - 2x + 3x - 6 = 6x$. Simplify: $x^2 + x - 6 = 6x$. Subtract $6x$ from both sides: $x^2 - 5x - 6 = 0$. This quadratic factors as $(x-6)(x+1) = 0$. Either $x = 6$ or $x = -1$. Check the answers by plugging them into the original equation: $\frac{6+3}{6} = \frac{9}{6} = \frac{3}{2}$ agrees with $\frac{6}{6-2} = \frac{6}{4} = \frac{3}{2}$ and $\frac{-1+3}{6} = \frac{2}{6} = \frac{1}{3}$ agrees with $\frac{-1}{-1-2} = \frac{-1}{-3} = \frac{1}{3}$.

Example 1. $\frac{x}{3x^2-10} = \frac{2}{7}$. Cross multiply: $7x = 6x^2 - 20$. Subtract $7x$ from both sides: $0 = 6x^2 - 7x - 20$. You could use the quadratic formula, but this equation happens to be simpler to factor: $0 = (3x+4)(2x-5)$. Either $3x + 4 = 0$ or $2x - 5 = 0$, which means that either $x = -\frac{4}{3}$ or $x = \frac{5}{2}$.

Check: $\frac{-4/3}{3(-4/3)^2-10} = \frac{-4/3}{3\left(\frac{16}{9}\right)-10} = \frac{-4/3}{\frac{16}{3}-\frac{30}{3}} = \frac{-4/3}{-14/3} = \frac{4}{3} \div \frac{14}{3} = \frac{4}{3}\cdot\frac{3}{14} = \frac{4}{14} = \frac{2}{7}$ and $\frac{5/2}{3(5/2)^2-10} = \frac{5/2}{3\left(\frac{25}{4}\right)-10} = \frac{5/2}{\frac{75}{4}-\frac{40}{4}} = \frac{5/2}{35/4} = \frac{5}{2} \div \frac{35}{4} = \frac{5}{2}\cdot\frac{4}{35} = \frac{20}{70} = \frac{2}{7}$.

4.7 Problems

Directions: Cross multiply and then solve the resulting quadratic equation.

(1) $\frac{x}{4} = \frac{x+8}{x}$

(2) $\frac{x^2+2}{x} = \frac{19}{3}$

(3) $\dfrac{8}{x+7} = -\dfrac{2x}{3}$

(4) $\dfrac{x+3}{3} = \dfrac{14}{x+2}$

(5) $\dfrac{2}{9x} = \dfrac{5}{3-12x^2}$

(6) $\dfrac{3x}{4x-3} = \dfrac{2}{x-1}$

(7) $\dfrac{x}{2x^2-15} = \dfrac{3}{17}$

(8) $\dfrac{1}{8x+7} = -\dfrac{4x}{3}$

(9) $\dfrac{x-3}{6} = \dfrac{2}{x-3}$

(10) $\dfrac{x}{29} = \dfrac{1}{10-x}$

5 SYSTEMS

5.1 Substitution

One way to solve a system of equations is to first solve for one unknown in terms of the other and then plug this into the unused equation(s), as in the examples.

Example 1. $5x + 4y = 42$ and $2x + 8y = 36$.
Step 1: Isolate x in the second equation.
Subtract $8y$ in the second equation: $2x = 36 - 8y$. Simplify: $x = 18 - 4y$.
Step 2: Use this equation to eliminate x from the first equation.
Plug $x = 18 - 4y$ into the first equation: $5(18 - 4y) + 4y = 42$.
Distribute: $90 - 20y + 4y = 42$. Simplify: $90 - 16y = 42$. Add $16y$ and subtract 42:
$48 = 16y$. Divide by 16 on both sides: $3 = y$.
Step 3: Plug $y = 3$ into the equation $x = 18 - 4y$ from earlier.
$x = 18 - 4(3) = 18 - 12 = 6$.
Check: $5(6) + 4(3) = 30 + 12 = 42$ and $2(6) + 8(3) = 12 + 24 = 36$.

Example 2. $2x - 5y + 6z = 5$, $5x + 3y - 9z = 11$, and $9x - 2y - 3z = 24$.
Step 1: Isolate x in the first equation.
$2x = 5 + 5y - 6z$. Divide by 2: $x = \frac{5}{2} + \frac{5y}{2} - 3z$.
Step 2: Plug x into each of the other two equations.
$5\left(\frac{5}{2} + \frac{5y}{2} - 3z\right) + 3y - 9z = 11$ and $9\left(\frac{5}{2} + \frac{5y}{2} - 3z\right) - 2y - 3z = 24$. Distribute:
$\frac{25}{2} + \frac{25y}{2} - 15z + 3y - 9z = 11$ and $\frac{45}{2} + \frac{45y}{2} - 27z - 2y - 3z = 24$.
Tip: Multiply each equation by the LCD (which is 2) to eliminate the fractions.
$25 + 25y - 30z + 6y - 18z = 22$ and $45 + 45y - 54z - 4y - 6z = 48$. Simplify:
$25 + 31y - 48z = 22$ and $45 + 41y - 60z = 48$.
Step 3: Isolate y in the equation $25 + 31y - 48z = 22$.
$31y = 48z - 3$. Divide by 31: $y = \frac{48z}{31} - \frac{3}{31}$.
Step 4: Plug y into $45 + 41y - 60z = 48$.

$45 + 41\left(\frac{48z}{31} - \frac{3}{31}\right) - 60z = 48$. Distribute: $45 + \frac{1968z}{31} - \frac{123}{31} - 60z = 48$. Multiply by 31: $1395 + 1968z - 123 - 1860z = 1488$. Simplify: $1272 + 108z = 1488$. Subtract 1272 from both sides: $108z = 216$. Divide by 108 on both sides: $z = \frac{216}{102} = 2$.

Step 5: Plug $z = 2$ into $y = 48z/31 - 3/31$ from Step 3.

$y = \frac{48(2)}{31} - \frac{3}{31} = \frac{96}{31} - \frac{3}{31} = \frac{93}{31} = 3$.

Step 6: Plug $y = 3$ and $z = 2$ into $x = 5/2 + 5y/2 - 3z$ from Step 1.

$x = \frac{5}{2} + \frac{5(3)}{2} - 3(2) = \frac{5}{2} + \frac{15}{2} - 6 = \frac{20}{2} - 6 = 10 - 6 = 4$.

Check: $2(4) - 5(3) + 6(2) = 8 - 15 + 12 = 5, 5(4) + 3(3) - 9(2) = 20 + 9 - 18 = 11$, and $9(4) - 2(3) - 3(2) = 36 - 6 - 6 = 24$.

5.1 Problems

Directions: Solve each system of equations for all of the unknowns.

(1) $8x - 4y = 20$
     $2x + 5y = 59$

(2) $5x + 3y = 16$
     $4x + 7y = 68$

(3) $4x - 9y = 52$
     $6x - 3y = -6$

(4) $-5x + 9y = 39$
     $2x - 8y = -42$

(5) $8x + 3y = 22$
$4x - 6y = -9$

(6) $5x - 6y = 8$
$7x + 9y = 17$

(7) $7x - 2y = -19$
$-x + 6y = 7$

(8) $8x + 4y = 8$
$2x + 9y = 12$

(9) $5x + 3y = 2$
$10x - 6y = 2$

(10) $-3x - 7y = 15$
$-7x + 3y = -52$

(11) $8x - 9y = 60$
$4x + 2y = -100$

(12) $9x - 3y = \frac{1}{2}$
$5x + 2y = \frac{10}{3}$

(13) $4x + 7y - 2z = 55$
$8x - 3y - 6z = 9$
$4y = 20$

(14) $3x - 8y + 5z = 27$
$9x - 4y - 7z = 11$
$6y + 3z = 27$

(15) $8x + 9y + 6z = 27$
$6x + 12y - 3z = 5$
$12x - 6y - 4z = -5$

(16) $10x - 15y + 20z = 71$
$25x + 5y - 10z = 8$
$15x - 30y - 5z = 27$

5.2 Simultaneous Equations

A pair of equations can often be solved more efficiently by operating on both of the equations simultaneously than by using the method of substitution. The main idea is to multiply both sides of each equation by a number with the goal of making **opposite coefficients** of the same variable in each equation, as illustrated in the examples.

Example 1. $5x - 6y = -32$ and $4x + 9y = 2$.
The 6 (in $-6y$) and the 9 (in $9y$) each evenly divide into 18. Multiply both sides of the left equation by 3 and multiply both sides of the right equation by 2 in order to make opposite coefficients $-18y$ and $18y$.
$$15x - 18y = -96 \quad \text{and} \quad 8x + 18y = 4$$
Add the two equations together. The sum of the left-hand sides equals the sum of the right-hand sides. In the sum of the left-hand sides, $18y$ cancels out.
$$15x - 18y + 8x + 18y = -96 + 4$$
$$23x = -92$$
$$x = -\frac{92}{23} = -4$$
Plug $x = -4$ into any of the previous equations to solve for y.
$$5(-4) - 6y = -32$$
$$-20 - 6y = -32$$
$$-6y = -12$$
$$y = \frac{-12}{-6} = 2$$
Check: $5(-4) - 6(2) = -20 - 12 = -32$ and $4(-4) + 9(2) = -16 + 18 = 2$.

Example 2. $5x + 9y = 52$ and $4x + 7y = 41$.
Multiply both sides of the left equation by 4 and multiply both sides of the right equation by -5 in order to make opposite coefficients $20x$ and $-20x$. **Every term of the right equation changes sign.**
$$20x + 36y = 208 \quad \text{and} \quad -20x - 35y = -205$$
Add the two equations together. On the left-hand side, $20x$ cancels out.

$$36y - 35y = 208 - 205$$
$$y = 3$$

Plug $y = 3$ into any of the previous equations to solve for x.

$$4x + 7(3) = 41$$
$$4x + 21 = 41$$
$$4x = 20$$
$$x = \frac{20}{4} = 5$$

Check: $5(5) + 9(3) = 25 + 27 = 52$ and $4(5) + 7(3) = 20 + 21 = 41$.

5.2 Problems

Directions: Solve each system of equations for all of the unknowns.

(1) $2x + 6y = 50$
      $7x - 8y = 30$

(2) $7x + 9y = 30$
      $3x + 5y = 22$

(3) $7x - 9y = 27$
      $4x + 3y = 48$

(4) $8x + 7y = 15$
      $-5x - 6y = 2$

(5) $8x - 12y = -60$
      $7x - 9y = -42$

(6) $15x - 8y = -6$
      $12x + 9y = 26$

(7) $4x + 8y = 4$
      $6x - 4y = 26$

(8) $-9x + 25y = 9$
      $12x + 35y = 29$

(9) $16x + 36y = 32$
      $12x + 21y = 17$

(10) $11x - 5y = -2$
      $15x + 7y = 18$

(11) $4x - 7y = -\frac{77}{12}$
      $8x - 5y = \frac{89}{12}$

(12) $-7x + 3y = -\frac{51}{8}$
      $2x - 9y = 12$

5.3 Determinants

When parentheses surround an array of numbers, this refers to a **matrix**. When two vertical lines surround an array of numbers, this means to find the **determinant** of the matrix. To find the determinant of a matrix that has 2 rows and 2 columns, multiply along the diagonals and subtract as indicated below.

$$\begin{vmatrix} a_1 & b_1 \\ a_2 & b_2 \end{vmatrix} = \det \begin{pmatrix} a_1 & b_1 \\ a_2 & b_2 \end{pmatrix} = a_1 b_2 - b_1 a_2$$

The letters a and b represent the columns and the subscripts 1 and 2 represent the rows. For example, a_2 is the **element** (which is a number) of the matrix in the second row of the first column. To find the determinant of a matrix that has 3 rows and 3 columns, multiply along the diagonals and subtract as indicated below.

$$\begin{vmatrix} a_1 & b_1 & c_1 \\ a_2 & b_2 & c_2 \\ a_3 & b_3 & c_3 \end{vmatrix} = a_1 b_2 c_3 + b_1 c_2 a_3 + c_1 a_2 b_3 - c_1 b_2 a_3 - b_1 a_2 c_3 - a_1 c_2 b_3$$

Example 1. $\begin{vmatrix} 3 & 5 \\ 4 & 8 \end{vmatrix} = 3(8) - 5(4) = 24 - 20 = 4$

Example 2. $\begin{vmatrix} 4 & 5 & 3 \\ 7 & 1 & 8 \\ 6 & 9 & 2 \end{vmatrix} = 4(1)(2) + 5(8)(6) + 3(7)(9) - 3(1)(6) - 5(7)(2) - 4(8)(9)$

$= 8 + 240 + 189 - 18 - 70 - 288 = 437 - 376 = 61$

5.3 Problems

Directions: Find each determinant.

(1)

$$\begin{vmatrix} 9 & 7 \\ 6 & 8 \end{vmatrix}$$

(2)

$$\begin{vmatrix} 5 & 8 \\ 3 & -4 \end{vmatrix}$$

(3)

$$\begin{vmatrix} 7 & 8 \\ -6 & -5 \end{vmatrix}$$

(4)

$$\begin{vmatrix} 9 & 4 \\ 4 & 9 \end{vmatrix}$$

(5)

$$\begin{vmatrix} 6 & 1 & 8 \\ 7 & 5 & 3 \\ 2 & 9 & 4 \end{vmatrix}$$

(6)

$$\begin{vmatrix} 1 & 2 & 3 \\ 2 & 4 & 5 \\ 3 & 5 & 6 \end{vmatrix}$$

(7)

$$\begin{vmatrix} 2 & 3 & 4 \\ 1 & -2 & -3 \\ 0 & -1 & -5 \end{vmatrix}$$

(8)

$$\begin{vmatrix} 4 & 3 & 3 \\ 3 & 4 & 0 \\ 3 & 0 & 4 \end{vmatrix}$$

(9)

$$\begin{vmatrix} 4 & -3 & 5 \\ -2 & 1 & 4 \\ 3 & 5 & 2 \end{vmatrix}$$

(10)

$$\begin{vmatrix} 2 & 25 & 10 \\ 25 & 5 & 20 \\ 2 & 20 & 10 \end{vmatrix}$$

(11)

$$\begin{vmatrix} 6 & 9 & 4 \\ 5 & 2 & 8 \\ 8 & 23 & -4 \end{vmatrix}$$

(12)

$$\begin{vmatrix} \frac{1}{2} & 12 & \frac{3}{2} \\ 8 & \frac{2}{3} & 4 \\ \frac{1}{4} & 6 & \frac{3}{4} \end{vmatrix}$$

5.4 Cramer's Rule

Cramer's rule provides formulas for solving a system of linear equations that have the following form:

$$a_1 x + b_1 y + c_1 z + \cdots = n_1$$
$$a_2 x + b_2 y + c_2 z + \cdots = n_2$$
$$a_3 x + b_3 y + c_3 z + \cdots = n_3$$

First, make the following matrices from the coefficients. The equations below apply when there are 3 equations with 3 unknowns, but the same idea applies if there are n equations with n unknowns (provided that they have the form shown above). The main idea is that the matrices for x, y, and z use the values of n_1, n_2, and n_3 to replace one column of the matrix of coefficients.

$$D_c = \begin{vmatrix} a_1 & b_1 & c_1 \\ a_2 & b_2 & c_2 \\ a_3 & b_3 & c_3 \end{vmatrix}, D_x = \begin{vmatrix} n_1 & b_1 & c_1 \\ n_2 & b_2 & c_2 \\ n_3 & b_3 & c_3 \end{vmatrix}, D_y = \begin{vmatrix} a_1 & n_1 & c_1 \\ a_2 & n_2 & c_2 \\ a_3 & n_3 & c_3 \end{vmatrix}, D_z = \begin{vmatrix} a_1 & b_1 & n_1 \\ a_2 & b_2 & n_2 \\ a_3 & b_3 & n_3 \end{vmatrix}$$

Cramer's rule gives the following solution to the system of equations.

$$x = \frac{D_x}{D_c} \quad , \quad y = \frac{D_y}{D_c} \quad , \quad z = \frac{D_z}{D_c}$$

Example 1. $5x + 6y = 11$ and $3x - 2y = 29$.

Identify $a_1 = 5, b_1 = 6, n_1 = 11, a_2 = 3, b_2 = -2$, and $n_2 = 29$.

$$D_c = \begin{vmatrix} a_1 & b_1 \\ a_2 & b_2 \end{vmatrix} = \begin{vmatrix} 5 & 6 \\ 3 & -2 \end{vmatrix} = 5(-2) - 6(3) = -10 - 18 = -28$$

$$D_x = \begin{vmatrix} n_1 & b_1 \\ n_2 & b_2 \end{vmatrix} = \begin{vmatrix} 11 & 6 \\ 29 & -2 \end{vmatrix} = 11(-2) - 6(29) = -22 - 174 = -196$$

$$D_y = \begin{vmatrix} a_1 & n_1 \\ a_2 & n_2 \end{vmatrix} = \begin{vmatrix} 5 & 11 \\ 3 & 29 \end{vmatrix} = 5(29) - 11(3) = 145 - 33 = 112$$

$$x = \frac{D_x}{D_c} = \frac{-196}{-28} = 7 \quad , \quad y = \frac{D_y}{D_c} = \frac{112}{-28} = -4$$

Check: $5(7) + 6(-4) = 35 - 24 = 11$ and $3(7) - 2(-4) = 21 + 8 = 29$.

Example 2. $3x - 2y + 6z = 3$, $6x + 7y - 4z = 56$, and $5x - 3y - 8z = -11$.

Identify $a_1 = 3$, $b_1 = -2$, $c_1 = 6$, $n_1 = 3$, $a_2 = 6$, $b_2 = 7$, $c_2 = -4$, $n_2 = 56$, and $a_2 = 5$, $b_3 = -3$, $c_3 = -8$, and $n_3 = -11$.

$$D_c = \begin{vmatrix} 3 & -2 & 6 \\ 6 & 7 & -4 \\ 5 & -3 & -8 \end{vmatrix} = 3(7)(-8) + (-2)(-4)(5) + 6(6)(-3) - 6(7)(5) - (-2)(6)(-8) - 3(-4)(-3)$$

$$= -168 + 40 - 108 - 210 - 96 - 36 = -578$$

$$D_x = \begin{vmatrix} 3 & -2 & 6 \\ 56 & 7 & -4 \\ -11 & -3 & -8 \end{vmatrix} = 3(7)(-8) - 2(-4)(-11) + 6(56)(-3) - 6(7)(-11) - (-2)(56)(-8) - 3(-4)(-3)$$

$$= -168 - 88 - 1008 + 462 - 896 - 36 = -1734$$

$$D_y = \begin{vmatrix} 3 & 3 & 6 \\ 6 & 56 & -4 \\ 5 & -11 & -8 \end{vmatrix} = 3(56)(-8) + 3(-4)(5) + 6(6)(-11) - 6(56)(5) - 3(6)(-8) - 3(-4)(-11)$$

$$= -1344 - 60 - 396 - 1680 + 144 - 132 = -3468$$

$$D_z = \begin{vmatrix} 3 & -2 & 3 \\ 6 & 7 & 56 \\ 5 & -3 & -11 \end{vmatrix} = 3(7)(-11) - 2(56)(5) + 3(6)(-3) - 3(7)(5) - (-2)(6)(-11) - 3(56)(-3)$$

$$= -231 - 560 - 54 - 105 - 132 + 504 = -578$$

$$x = \frac{D_x}{D_c} = \frac{-1734}{-578} = 3 \quad , \quad y = \frac{D_y}{D_c} = \frac{-3468}{-578} = 6 \quad , \quad z = \frac{D_z}{D_c} = \frac{-578}{-578} = 1$$

Check: $3(3) - 2(6) + 6(1) = 9 - 12 + 6 = 3$, $6(3) + 7(6) - 4(1) = 18 + 42 - 4 = 56$, and $5(3) - 3(6) - 8(1) = 15 - 18 - 8 = -11$.

5.4 Problems

Directions: Solve each system of equations for all of the unknowns.

(1) $5x + 4y = 7$
$9x + 8y = 19$

(2) $7x - 2y = 18$
$3x + 4y = 66$

(3) $8x - 6y = -12$
     $4x - 9y = -20$

(4) $7x + 4y = -6$
     $6x + 8y = 0$

(5) $2x + 8y = 28$
     $9x - 4y = 16$

(6) $9x + 2y = -2$
     $3x - 6y = 1$

(7) $8x + 5y - 4z = 3$
     $3x - y + 5z = 7$
     $-7x - 4y + 3z = -2$

(8) $4x - 9y = 2$
     $3y + 2z = 24$
     $8x - 5z = -5$

(9) $6x + 7y - 3z = 9$
$2x - 9y - 7z = 1$
$7x + 8y + 2z = 44$

(10) $4x - 6y + 9z = 20$
$7x - 8y - 8z = 60$
$3x + 4y - 6z = 32$

(11) $9x - 8y + 4z = 10$
$3x \quad\quad + 8z = 22$
$6x + 4y - 6z = -8$

(12) $5x + 7y + 2z = 9$
$8x - 5y - 4z = -12$
$4x - 2y + 9z = 2$

5.5 Nonlinear Systems

Many systems of nonlinear equations can be solved using the method of substitution, meaning to isolate one variable in one equation and then plug this into each of the unused equations (like we did in Sec. 5.1 for systems of linear equations).

Example 1. $\frac{y}{x} = 3$ and $\frac{1}{x} + \frac{1}{y} = \frac{1}{3}$.

Multiply by x on both sides of the first equation: $y = 3x$. Plug this into the second equation: $\frac{1}{x} + \frac{1}{3x} = \frac{1}{3}$. Tip: You can avoid dealing with fractions by multiplying both sides of the equation by the least common denominator (LCD). The LCD of x, $3x$, and 3 is $3x$, so multiply by $3x$ on both sides: $\frac{3x}{x} + \frac{3x}{3x} = \frac{3x}{3}$. Simplify: $3 + 1 = 4 = x$. Plug $x = 4$ into $y = 3x$ to get $y = 3(4) = 12$.

Check: $\frac{y}{x} = \frac{12}{4} = 3$ and $\frac{1}{4} + \frac{1}{12} = \frac{3}{12} + \frac{1}{12} = \frac{4}{12} = \frac{1}{3}$.

Example 2. $y + \sqrt{x} = 22$ and $x - y = 20$.

Isolate the radical in the first equation: $\sqrt{x} = 22 - y$. Square both sides: $x = (22 - y)^2$. Apply the FOIL method: $x = 484 - 44y + y^2$. Substitute this into $x - y = 20$: $484 - 44y + y^2 - y = 20$. Combine like terms and reorder: $y^2 - 45y + 464 = 0$. Now that the equation is in standard form, we may use the quadratic formula (Chapter 4). Note that $a = 1$, $b = -45$, and $c = 464$:

$$y = \frac{-(-45) \pm \sqrt{(-45)^2 - 4(1)(464)}}{2(1)} = \frac{45 \pm \sqrt{2025 - 1856}}{2} = \frac{45 \pm \sqrt{169}}{2} = \frac{45 \pm 13}{2}$$

$$y = \frac{45 + 13}{2} = \frac{58}{2} = 29 \quad \text{or} \quad y = \frac{45 - 13}{2} = \frac{32}{2} = 16$$

Plug each value into $x - y = 20$ to solve for x: $x - 29 = 20$ or $x - 16 = 20$. Simplify: $x = 29 + 20 = 49$ or $x = 16 + 20 = 36$.

There are two possibilities. Either $x = 36$ and $y = 16$ or $x = 49$ and $y = 29$. Only one of these turns out to correctly solve both of the original equations. The only correct answer is $x = 36$ and $y = 16$.

Check: $16 + \sqrt{36} = 16 + 6 = 22$ and $36 - 16 = 20$. (If you try the other possibility, you will see that it doesn't satisfy the equation $y + \sqrt{x} = 22$.)

5.5 Problems

Directions: Solve each system of equations for all of the unknowns.

(1) $3xy = 108$

  $\dfrac{2y}{x} = 288$

(2) $x + y = 27$

  $xy = 180$

(3) $y - x = 9$

  $\dfrac{1}{x} + \dfrac{1}{y} = \dfrac{1}{6}$

(4) $\dfrac{18}{x} = \dfrac{y}{8}$

  $x - 6 = 3y$

(5) $x\sqrt{y} = 36$

$y\sqrt{x} = 48$

(6) $3x^2 - 2y^2 = 19$

$x + y = 15$

(7) $x - \sqrt{y} = 7$

$x^2 + y = 169$

(8) $4x + 7y = 11$

$x^2 + y^2 = 73$

6 POLYNOMIALS

6.1 Multiply Polynomials

A **polynomial** is an expression where each term has the form ax^n. The highest power of the polynomial is called the **degree**. For example, $2x^4 + 7x^3 - x^2 + 6$ is degree four and $x^5 - 3x^2$ is degree five. The number that multiplies the variable is a **coefficient**. For example, 8 and 3 are coefficients in $8x^2 + 3x$.

Two polynomials can be multiplied by multiplying each term of one polynomial by each term of the other polynomial.

Note: The method below organizes the terms by their powers. This makes it easy to find the coefficient of each term in the answer. For example, $-2x^2 - 12x^2 + 5x^2 = -9x^2$ in the solution to Example 1 and $2x^2 - 18x^2 = -16x^2$ in the solution to Example 2.

Example 1. $(x^2 - 4x + 5)(x^2 + 3x - 2)$

$= x^4 + 3x^3 - 2x^2$
$\quad\quad - 4x^3 - 12x^2 + 8x$
$\quad\quad\quad\quad + 5x^2 + 15x - 10$
$= x^4 - x^3 - 9x^2 + 23x - 10$

Example 2. $(2x^2 - 6)(x^3 + 3x^2 - 4x + 1)$

$= 2x^5 + 6x^4 - 8x^3 + 2x^2$
$\quad\quad\quad - 6x^3 - 18x^2 + 24x - 6$
$= 2x^5 + 6x^4 - 14x^3 - 16x^2 + 24x - 6$

6.1 Problems

Directions: Multiply the polynomials together and combine like terms in the answer.

(1) $(5x + 2)(3x^2 - 6x + 4)$

(2) $(x - 1)(2x^2 + 8x - 5)$

(3) $(x^2 + 6x + 3)(x^2 + 4x + 7)$

(4) $(3x^2 + 8x - 5)(4x^2 - 2x + 9)$

(5) $(2x^2 - 5x + 6)(5x^2 + 7x - 4)$

(6) $(6x^2 - x - 8)^2$

(7) $(x + 3)(4x^2 + 9x - 7)$

(8) $(2x - 5)(3x^3 - 8x^2 + 5x - 2)$

(9) $(3x^2 + 4)(2x^3 + 6x^2 - 7x + 9)$

(10) $(8x^2 - 7x)(5x^6 - 4x^4 - 8x^2 + 3)$

6.2 Polynomial Division

In the division problem $20 \div 4 = 5$, which can be expressed as $\frac{20}{4} = 5$, the number 20 is the **dividend**, the number 4 is the **divisor**, and the number 5 is the **quotient**. Note that the fraction $\frac{20}{4}$, which is equivalent to 5, may also be referred to as the quotient.

Consider $8x^3 + 24x^2 + 8x - 15$ divided by $2x + 3$. Since the highest powers of the dividend and divisor are x^3 and x, the highest power of the quotient is $\frac{x^3}{x} = x^2$. Since the lowest powers of the dividend and divisor are constants, the lowest power of the quotient is a constant. This means that the quotient has the form $ax^2 + bx + c$. This division problem can be expressed as $\frac{8x^3+24x^2+8x-15}{2x+3} = ax^2 + bx + c$. To solve this division problem, we need to determine a, b, and c.

One way to find a, b, and c is to multiply by $2x + 3$ on both sides of $\frac{8x^3+24x^2+8x-15}{2x+3} = ax^2 + bx + c$. This gives $8x^3 + 24x^2 + 8x - 15 = (2x + 3)(ax^2 + bx + c)$. Multiply: $8x^3 + 24x^2 + 8x - 15 = 2ax^3 + 2bx^2 + 2cx + 3ax^2 + 3bx + 3c$. Combine like terms: $8x^3 + 24x^2 + 8x - 15 = 2ax^3 + (3a + 2b)x^2 + (3b + 2c)x + 3c$. In order for both sides of this equation to be equal for all possible values of x, the coefficients on both sides must match up. Setting corresponding coefficients equal, we get four equations:

$$8 = 2a$$
$$24 = 3a + 2b$$
$$8 = 3b + 2c$$
$$-15 = 3c$$

The top equation gives $a = 4$ and the bottom equation gives $c = -5$. Plug $a = 4$ into the second equation: $24 = 3(4) + 2b$. Simplify: $12 = 2b$, such that $b = 6$. Check the third equation for consistency: $3b + 2c = 3(6) + 2(-5) = 18 - 10 = 8$. The quotient is therefore $ax^2 + bx + c = 4x^2 + 6x - 5$. Check the answer by multiplying:

$$(2x + 3)(4x^2 + 6x - 5) = 8x^3 + 12x^2 - 10x + 12x^2 + 18x - 15$$
$$= 8x^3 + 24x^2 + 8x - 15$$

Another way to divide two polynomials is to use the method of **long division**. This is very similar to long division for arithmetic, except for using polynomials.

1. Multiply $2x + 3$ by $4x^2$ to make $8x^3$.
2. $(2x + 3)4x^2 = 8x^3 + 12x^2$.
3. Subtract $8x^3 + 12x^2$ from the dividend.
4. Bring down the $8x$ from the dividend.
5. Multiply $2x + 3$ by $6x$ to make $12x^2$.
6. $(2x + 3)6x = 12x^2 + 18x$.
7. Subtract $12x^2 + 18x$ from the dividend.
8. Bring down the -15 from the dividend.
9. Multiply $2x + 3$ by -5 to make $-10x$.
10. The remainder is zero. The answer is $4x^2 + 6x - 5$.

$$
\begin{array}{r}
4x^2 + 6x - 5 \\
2x + 3 \,\overline{\big)\, 8x^3 + 24x^2 + 8x - 15} \\
\underline{8x^3 + 12x^2} \\
12x^2 + 8x \\
\underline{12x^2 + 18x} \\
-10x - 15 \\
\underline{-10x - 15} \\
0
\end{array}
$$

Example 1. $\dfrac{20x^3 - 38x^2 + 27x - 6}{5x - 2}$

$$
\begin{array}{r}
4x^2 - 6x + 3 \\
5x - 2 \,\overline{\big)\, 20x^3 - 38x^2 + 27x - 6} \\
\underline{20x^3 - 8x^2} \\
-30x^2 + 27x \\
\underline{-30x^2 + 12x} \\
15x - 6 \\
\underline{15x - 6} \\
0
\end{array}
$$

Example 2. $\dfrac{10x^4 + 11x^3 - 18x^2 + 20x - 16}{2x^2 + 3x - 4}$

$$
\begin{array}{r}
5x^2 - 2x + 4 \\
2x^2 + 3x - 4 \,\overline{\big)\, 10x^4 + 11x^3 - 18x^2 + 20x - 16} \\
\underline{10x^4 + 15x^3 - 20x^2} \\
-4x^3 + 2x^2 + 20x \\
\underline{-4x^3 - 6x^2 + 8x} \\
8x^2 + 12x - 16 \\
\underline{8x^2 + 12x - 16} \\
0
\end{array}
$$

Notes: First multiply $4x^2$ by $5x - 2$ in Ex. 1 and multiply $5x^2$ by $2x^2 + 3x - 4$ in Ex. 2. It's easy to make mistakes when subtracting with minus signs:

- $-38x^2 - (-8x^2) = -38x^2 + 8x^2 = -30x^2$ in Example 1.
- $-18x^2 - (-20x^2) = -18x^2 + 20x^2 = 2x^2$ in Example 2.
- $2x^2 - (-6x)^2 = 2x^2 + 6x^2 = 8x^2$ in Example 2.

An alternative to long division is to solve the problems like we did on the previous page. Check for Example 1: $(5x - 2)(4x^2 - 6x + 3) = 20x^3 - 30x^2 + 15x - 8x^2 + 12x - 6 = 20x^3 - 38x^2 + 27x - 6$.

Check for Example 2: $(2x^2 + 3x - 4)(5x^2 - 2x + 4) = 10x^4 - 4x^3 + 8x^2 + 15x^3 - 6x^2 + 12x - 20x^2 + 8x - 16 = 10x^4 + 11x^3 - 18x^2 + 20x - 16$.

6.2 Problems

Directions: Divide the polynomials.

(1) $\dfrac{x^3+10x^2+29x+20}{x+4}$

(2) $\dfrac{4x^3-21x^2-25x+42}{x-6}$

(3) $\dfrac{18x^5-48x^4-27x^3+72x^2+15x-40}{3x-8}$

(4) $\dfrac{10x^4+46x^3-11x^2-81x-36}{5x+3}$

(5) $\dfrac{4x^4+32x^3+19x^2+56x+21}{4x^2+7}$

(6) $\dfrac{4x^4+18x^3+15x^2+37x+36}{x^2+5x+4}$

(7) $\dfrac{6x^4+26x^3-5x^2-62x+35}{2x^2+6x-5}$

(8) $\dfrac{8x^5+6x^4-86x^3+176x^2-141x+40}{4x^2-9x+8}$

6.3 Remainders

Just as in arithmetic, when dividing polynomials, sometimes the answer includes a remainder. In ordinary arithmetic, the answer to $20 \div 3$ equals $6 + \frac{2}{3}$. In this example, the remainder is 2. Check the answer by multiplying and applying the distributive property: $3\left(6 + \frac{2}{3}\right) = 18 + 3\left(\frac{2}{3}\right) = 18 + 2 = 20$. Note that we refer to the remainder as 2, even though the answer includes the fraction $\frac{2}{3}$. The remainder is less than the divisor. For example, in $20 \div 3 = 6 + \frac{2}{3}$ the divisor is 3 and the remainder is 2. When the polynomial division $\frac{f(x)}{g(x)}$ has a remainder (recall the notation for functions from Chapter 1), the answer can be expressed as $q(x) + \frac{r(x)}{g(x)}$, where $q(x)$ is the **quotient**, $f(x)$ is the **dividend**, $g(x)$ is the **divisor**, and $r(x)$ is the **remainder**. To be consistent with the language used in ordinary arithmetic, we refer to $r(x)$ as the remainder, even though the answer includes the fraction $\frac{r(x)}{g(x)}$. The degree of the remainder is **less than** the degree of the divisor, meaning that the highest power of the remainder is less than the highest power of the divisor.

Example 1. $\dfrac{6x^3+23x^2+18x+7}{2x+5}$

$$
\begin{array}{r}
3x^2 + 4x - 1 \\
2x + 5 \overline{\smash{\big)}\, 6x^3 + 23x^2 + 18x + 7} \\
\underline{6x^3 + 15x^2} \\
8x^2 + 18x \\
\underline{8x^2 + 20x} \\
- 2x + 7 \\
\underline{- 2x - 5} \\
12
\end{array}
$$

Example 2. $\dfrac{9x^4-3x^3+4x^2+2x-5}{3x^2-2x}$

$$
\begin{array}{r}
3x^2 + x + 2 \\
3x^2 - 2x \overline{\smash{\big)}\, 9x^4 - 3x^3 + 4x^2 + 2x - 5} \\
\underline{9x^4 - 6x^3} \\
3x^3 + 4x^2 \\
\underline{3x^3 - 2x^2} \\
6x^2 + 2x \\
\underline{6x^2 - 4x} \\
6x - 5
\end{array}
$$

$\dfrac{6x^3+23x^2+18x+7}{2x+5} = 3x^2 + 4x - 1 + \dfrac{12}{2x+5}$

$\dfrac{9x^4-3x^3+4x^2+2x-5}{3x^2-2x} = 3x^2 + x + 2 + \dfrac{6x-5}{3x^2-2x}$

It's easy to make mistakes when subtracting with minus signs:

- $7 - (-5) = 12$ in Example 1.
- $-3x^3 - (-6x^3) = -3x^3 + 6x^3 = 3x^3$ in Example 2.
- $4x^2 - (-2x)^2 = 4x^2 + 2x^2 = 6x^2$ and $2x - (-4x) = 2x + 4x = 6x$ in Example 2.

Check for Example 1: $(2x + 5)\left(3x^2 + 4x - 1 + \frac{12}{2x+5}\right) = (2x + 5)(3x^2 + 4x - 1) + 12 =$
$6x^3 + 8x^2 - 2x + 15x^2 + 20x - 5 + 12 = 6x^3 + 23x^2 + 18x + 7.$

Check for Example 2: $(3x^2 - 2x)\left(3x^2 + x + 2 + \frac{6x-5}{3x^2-2x}\right) = (3x^2 - 2x)(3x^2 + x + 2) +$
$6x - 5 = 9x^4 + 3x^3 + 6x^2 - 6x^3 - 2x^2 - 4x + 6x - 5 = 9x^4 - 3x^3 + 4x^2 + 2x - 5.$

6.3 Problems

Directions: Divide the polynomials.

(1) $\dfrac{x^3 + 11x^2 + 38x + 25}{x+6}$

(2) $\dfrac{24x^3 - 34x^2 - 34x + 50}{4x-7}$

(3) $\dfrac{18x^4 - 57x^3 + 65x^2 + 23x - 8}{6x-5}$

(4) $\dfrac{21x^4 + 68x^3 + 5x^2 - 78x - 40}{3x+8}$

(5) $\dfrac{36x^5+81x^4-2x^3+36x^2}{9x^2+4}$

(6) $\dfrac{30x^4+28x^3-36x^2+3x+9}{5x^2-2x}$

(7) $\dfrac{5x^5+13x^4-70x^3+69x^2-51x+33}{x^2+4x-9}$

(8) $\dfrac{6x^5-59x^3+85x^2-65x+35}{2x^2-6x+3}$

6.4 Synthetic Division

Synthetic division is a variation of long division that offers a more concise solution when dividing by a binomial of the form $x - c$ (where the coefficient of the variable is one and the variable is not raised to a power). There is one important detail to note: **The constant term is being subtracted.** This helps to avoid sign mistakes during the division process, but it will backfire if you forget that the constant is subtracted. For example, when dividing by $x - 8$, c is positive 8, but when dividing by $x + 6$, c is -6.

Synthetic division isn't really different from ordinary division. It is just a more concise way of writing the solution:

- Write the constant c to the left instead of the divisor $x - c$. Note the sign. For example, to divide by $x - 5$, write positive 5 to the left, and to divide by $x + 3$, write -3 to the left. **The sign is opposite.**
- Write the coefficients of the dividend on the top row.
- Repeat the first coefficient in the third row below the first coefficient.
- Multiply c by the first coefficient. Write the answer below the second coefficient.
- **Add** the second coefficient to the number below it. (In synthetic division, we add these numbers, whereas in ordinary division we subtracted. The reason is that in synthetic division, we work with the opposite sign as noted in the first bullet point above. This allows us to add numbers instead of subtracting them, which makes it easier to deal with minus signs.) Write this sum in the third row below the two numbers that you added.
- Multiply c by the second number in the bottom row. Write the answer below the third coefficient. Add the second coefficient to the number below it and write this number in the third row below these two numbers. Repeat this step until you reach the last column.
- If the last column of the third row isn't zero, there is a remainder. Draw a vertical bar to the left of the remainder.

The easiest way to learn synthetic division is to compare it with ordinary division in the examples. As you study the examples, return to this page and reread the directions as needed. The directions will make more sense when you study the examples.

Note: Ordinary division is shown on the left and synthetic division is shown on the right.

Example 1. $\dfrac{5x^3-7x^2-14x+22}{x-2}$ Note: $c = 2$ is positive. It is opposite to the -2 in $x - 2$.

$$
\begin{array}{r}
5x^2 + 3x\ -8 \\
x - 2\ \overline{\big)\ 5x^3 - 7x^2 - 14x + 22} \\
\underline{5x^3 - 10x^2} \\
3x^2 - 14x \\
\underline{3x^2 - 6x} \\
-8x + 22 \\
\underline{-8x + 16} \\
6
\end{array}
$$

$$
\begin{array}{c|rrrr}
2 & 5 & -7 & -14 & 22 \\
 & & 10 & 6 & -16 \\
\hline
 & 5 & 3 & -8\ & 6
\end{array}
$$

The answer is $\dfrac{5x^3-7x^2-14x+22}{x-2} = 5x^2 + 3x - 8 + \dfrac{6}{x-2}$.

Compare: Subtract in ordinary division (left), but **add** in synthetic division (right).

Notes: $2 \times 5 = 10$, $2 \times 3 = 6$, and $2 \times (-8) = -16$.

Check for Example 1: $(x - 2)\left(5x^2 + 3x - 8 + \dfrac{6}{x-2}\right) = (x - 2)(5x^2 + 3x - 8) + 6 =$

$5x^3 + 3x^2 - 8x - 10x^2 - 6x + 16 + 6 = 5x^3 - 7x^2 - 14x + 22$.

Example 2. $\dfrac{2x^4+14x^3+21x^2-7x+20}{x+4}$ Note: $c = -4$ is negative. It is opposite to the 4 in $x + 4$.

$$
\begin{array}{r}
2x^3 + 6x^2\ -3x\ +5 \\
x + 4\ \overline{\big)\ 2x^4 + 14x^3 + 21x^2 - 7x + 20} \\
\underline{2x^4 + 8x^3} \\
6x^3\ +21x^2 \\
\underline{6x^3\ +24x^2} \\
-3x^2 - 7x \\
\underline{-3x^2 - 12x} \\
5x + 20 \\
\underline{5x + 20} \\
0
\end{array}
$$

$$
\begin{array}{c|rrrrr}
-4 & 2 & 14 & 21 & -7 & 20 \\
 & & -8 & -24 & 12 & -20 \\
\hline
 & 2 & 6 & -3 & 5\ & 0
\end{array}
$$

The answer is $\dfrac{2x^4+14x^3+21x^2-7x+20}{x+4} = 2x^3 + 6x^2 - 3x + 5$.

Notes: $-4 \times 2 = -8$, $14 + (-8) = 6$, $-4 \times 6 = -24$, $21 + (-24) = -3$, $-4 \times (-3) =$
12, $-7 + 12 = 5$, $-4 \times 5 = -20$, and $20 + (-20) = 0$.

Check for Example 2: $(x + 4)(2x^3 + 6x^2 - 3x + 5) = 2x^4 + 6x^3 - 3x^2 + 5x + 8x^3 +$
$24x^2 - 12x + 20 = 2x^4 + 14x^3 + 21x^2 - 7x + 20$.

6.4 Problems

Directions: Use synthetic division to divide the polynomials.

(1) $\dfrac{x^3+6x^2-34x+23}{x-3}$

(2) $\dfrac{7x^3-51x^2-36x-32}{x-8}$

(3) $\dfrac{6x^3+26x^2-11x+44}{x+5}$

(4) $\dfrac{3x^4-17x^3-14x^2+43x+37}{x-6}$

(5) $\dfrac{5x^4-52x^3+65x^2-18x+5}{x-9}$

(6) $\dfrac{8x^4+22x^3+3x^2-14x+8}{x+2}$

(7) $\dfrac{2x^5+19x^4+32x^3-20x^2-53}{x+7}$

(8) $\dfrac{9x^5-43x^4+28x^3+9x^2-40x+16}{x-4}$

6.5 Synthetic Division with a Coefficient

To divide a polynomial by a binomial with the form $ax - b$, synthetic division can be used if you factor out the a like the example. At the end, remember to put the answer to the synthetic division back into the expression that you made by factoring.

Example 1. $\dfrac{6x^3-23x^2+29x-5}{3x-4} = \dfrac{6x^3-23x^2+29x-5}{3\left(x-\frac{4}{3}\right)} = \dfrac{1}{3}\dfrac{6x^3-23x^2+29x-5}{x-\frac{4}{3}}$ Note: $c = \dfrac{4}{3}$ is positive.

$$
\begin{array}{c|cccc}
\frac{4}{3} & 6 & -23 & 29 & -5 \\
 & & 8 & -20 & 12 \\
\hline
 & 6 & -15 & 9 & 7 \\
\end{array}
$$

Remember the $\frac{1}{3}$ from the factoring. Plug $6x^2 - 15x + 9 + \dfrac{7}{x-\frac{4}{3}}$ into $\dfrac{1}{3}\dfrac{6x^3-23x^2+29x-5}{x-\frac{4}{3}}$:

$$\frac{1}{3}\frac{6x^3-23x^2+29x-5}{x-\frac{4}{3}} = \frac{1}{3}\left(6x^2 - 15x + 9 + \frac{7}{x-\frac{4}{3}}\right) = \boxed{2x^2 - 5x + 3 + \frac{7}{3x-4}}.$$

Notes: $\frac{4}{3} \times 6 = \frac{24}{3} = 8, -23 + 8 = -15, \frac{4}{3} \times (-15) = -\frac{60}{3} = -20, 29 + (-20) = 9,$

$\frac{4}{3} \times 9 = \frac{36}{3} = 12,$ and $-5 + 12 = 7.$

Check for Example 1: $(3x - 4)\left(2x^2 - 5x + 3 + \dfrac{7}{3x-4}\right) = (3x - 4)(2x^2 - 5x + 3) + 7$

$= 6x^3 - 15x^2 + 9x - 8x^2 + 20x - 12 + 7 = 6x^3 - 23x^2 + 29x - 5.$

6.5 Problems

Directions: Use factoring and synthetic division to divide the polynomials.

(1) $\dfrac{14x^2-34x-19}{2x-6}$

(2) $\dfrac{8x^3-28x^2+56x-64}{4x-8}$

(3) $\dfrac{20x^3+17x^2-27x-22}{5x+3}$

(4) $\dfrac{18x^3-27x^2+42x-61}{6x-9}$

(5) $\dfrac{56x^4-67x^3-43x^2+64x-16}{7x-4}$

(6) $\dfrac{40x^4+74x^3+20x+5}{8x+2}$

(7) $\dfrac{9x^5+66x^4-84x^3+51x^2-72x+39}{9x-6}$

(8) $\dfrac{-6x^4+26x^3-33x^2+34x-21}{3-x}$

6.6 The Remainder Theorem

According to the **remainder theorem**, if $p(x)$ is a polynomial, the remainder of $\frac{p(x)}{x-c}$ is equal to $p(c)$. Recall from Sec. 1.3 that $p(c)$ means to evaluate the polynomial function $p(x)$ when x equals c.

Why does the remainder of $\frac{p(x)}{x-c}$ equals $p(c)$? First write $\frac{p(x)}{x-c} = q(x) + \frac{r(x)}{x-c}$, which states that the division $\frac{p(x)}{x-c}$ has a quotient $q(x)$ and remainder $r(x)$ (review Sec. 6.3). Multiply by $x - c$ on both sides: $p(x) = (x - c)q(x) + r(x)$. The remainder for $\frac{p(x)}{x-c}$ must be a constant because the degree of the remainder must be less than the degree of the divisor. Let $r(x) = d$ be the constant. Then $p(x) = (x - c)q(x) + d$. Plug $x = c$ into this to get $p(c) = (c - c)q(c) + d = 0 + d = d$. Since $r(x) = d$ and $p(c) = d$, this shows that $p(c)$ equals the remainder.

If a polynomial is divided by $ax - b$ instead of $x - c$, the remainder equals $p\left(\frac{b}{a}\right)$.

Example 1. Determine the remainder for $\frac{x^3+2x^2-21x+22}{x-3}$.
Identify $p(x) = x^3 + 2x^2 - 21x + 22$ and $c = 3$. Compare $x - 3$ with $x - c$ to see that c is positive. Calculate $p(3)$ like we did in Sec. 1.3.
$$p(3) = (3)^3 + 2(3)^2 - 21(3) + 22 = 27 + 2(9) - 63 + 22 = 49 + 18 - 63 = 4$$

Example 2. Determine the remainder for $\frac{x^5}{x+5}$.
Identify $p(x) = x^5$ and $c = -5$. Compare $x + 5$ with $x - c$ to see that c is negative.
$$p(-5) = (-5)^5 = (-5)(-5)(-5)(-5)(-5) = -3125$$

Example 3. Determine the remainder for $\frac{6x^2+13x-23}{3x-4}$.
Since the divisor has a coefficient, we will work with a and b (rather than c). Identify $p(x) = 6x^2 + 13x - 23$, $a = 3$, and $b = 4$. Compare $3x - 4$ with $ax - b$ to see that $b = 4$.
$$p\left(\frac{4}{3}\right) = 6\left(\frac{4}{3}\right)^2 + 13\left(\frac{4}{3}\right) - 23 = \frac{32}{3} + \frac{52}{3} - 23 = \frac{84}{3} - 23 = 28 - 23 = 5$$

6.6 Problems

Directions: Use the remainder theorem to determine the remainder.

(1) $\dfrac{5x-7}{x-4}$

(2) $\dfrac{x^2-9x+14}{x-7}$

(3) $\dfrac{2x^3+8x^2-7x+52}{x+6}$

(4) $\dfrac{5x^5-3x^4+6x^3-7x^2+9x-15}{x+1}$

(5) $\dfrac{8x^2+6x-5}{4x-1}$

(6) $\dfrac{6x^3+4x^2-3x+7}{2x+3}$

(7) $\dfrac{16x^5+64x^4-80x^3+48x^2-72x+39}{2x-5}$

(8) $\dfrac{4x^4}{6x-3}$

(9) $\dfrac{x^5-3x^4-5x^3+6x^2-7x+10}{4x+8}$

(10) $\dfrac{8x^3+3x+19}{4x+5}$

6.7 Factors and Roots of Polynomials

According to the **factor theorem**, if a polynomial $p(x)$ is evenly divisible by $x - c$, then $p(c) = 0$ (since the remainder must be zero), in which case $x - c$ is a factor of $p(x)$. The number c is a **root** of $p(x)$ if $p(c) = 0$, and may alternatively be called a **zero** of $p(x)$. The same strategies for finding roots (or zeroes) of a polynomial may thus be applied to factor a polynomial.

According to the **fundamental theorem of algebra**, if the degree of polynomial $p(x)$ equals n (which means that its highest power is x^n), then $p(x)$ has exactly n roots. Note that some of the roots may be complex numbers or irrational numbers (or both). It is important to beware that some of the roots might **not** be **distinct**, which means that some of the roots might be identical. For example, $x^2 - 2x + 1$ factors as $(x - 1)^2$ and has the double root $x = 1$, meaning that both roots are $x = 1$. In this case, the root $x = 1$ has a **multiplicity** of 2 (which means that it is a **double root**).

A polynomial $p(x)$ can be formed from its roots $c_1, c_2, \ldots, c_n$ as follows:
$$p(x) = k(x - c_1)(x - c_2) \cdots (x - c_n)$$
where k is a constant and the ellipsis ($\cdots$) means "and so on." You should be able to see that when the right-hand side of the above equation is multiplied out, the leading term will have a power of x^n, which helps to understand the reasoning behind the **fundamental theorem of algebra** (but this justification doesn't serve as a formal proof). You can also see that $p(c_1) = p(c_2) = \cdots = p(c_n) = 0$ in the above equation, which agrees with the **factor theorem** (and therefore also the **remainder theorem**).

If x_1 and x_2 are real numbers and the polynomial $p(x)$ has different signs when it is evaluated at $p(x_1)$ and $p(x_2)$, then $p(x)$ has a root that lies in $x_1 \leq c \leq x_2$. This can be useful in the following way. If you evaluate $p(x)$ at a variety of values and find two values of x where $p(x)$ has opposite signs, then you know that one of the roots lies in between these values of x. (Some numerical methods for finding approximate roots by writing code are based on this simple idea.) First try values of x for which it is easy to evaluate $p(x)$, such as $x = 1$ and $x = -1$.

If a polynomial doesn't have a constant term, like $x^3 - 4x^2 + 9x$, one root is $x = 0$.

Sometimes, it is easy to restrict the possible values of x which could be a root of $p(x)$. For example, $x^3 + 7x^2 + 3x + 14$ can't have a real root that is positive. If x is positive (and real), all four terms will be positive. A negative value of x allows the first and third terms to be negative, which could potentially cancel the second and fourth terms. (However, if x is complex, the second term could be negative in that case.) The rules below can help you narrow down your search for the roots of a polynomial:

- If every coefficient has the same sign, there are no real positive roots.
- If the coefficients of the even powers (and the sign of the constant term) all have opposite signs compared to the coefficients of the odd powers, there are no real negative roots.

If c is a root of $p(x)$, then divide $p(x)$ by $x - c$ to get $\frac{p(x)}{x-c} = q(x)$. The new polynomial $q(x)$ will be simpler than $p(x)$, and the roots of $q(x)$ will also be roots of $p(x)$. Every time you find a root, this technique helps to simplify the search for additional roots.

If the coefficients of $p(x) = a_n x^n + a_{n-1} x^{n-1} + a_{n-2} x^{n-2} + \cdots + a_2 x^2 + a_1 x + a_0$ are **integers**, if root $x = c$ is also an integer, then a_0 will be evenly divisible by c (meaning that a_0 will be a multiple of c). For example, consider $x^2 - 5x + 6$, which easily factors as $(x - 3)(x - 2)$ using the method of Sec. 4.2. In this case, the roots are $x = 2$ and $x = 3$, and $a_0 = 6$ is evenly divisible by 2 and 3. As another example, $3x^2 - 17x + 10$ factors as $(3x - 2)(x - 5)$ using the method of Sec. 4.3. In this case, the roots are $x = \frac{2}{3}$ and $x = 5$, and $a_0 = 10$ is a multiple of the root that is an integer. This fact about a_0 helps to narrow down the possible values of x for which a root can be an integer.

If the coefficients of $p(x) = a_n x^n + a_{n-1} x^{n-1} + a_{n-2} x^{n-2} + \cdots + a_2 x^2 + a_1 x + a_0$ are **integers**, if the **reduced fraction** $\frac{N}{D}$ is a root (and if N and D are each integers), then a_n will be evenly divisible by D and a_0 will be evenly divisible by N (meaning that a_n will be a multiple of the denominator, D, and a_0 will be a multiple of the numerator, N). For example, $12x^2 - 23x + 10$ factors as $(3x - 2)(4x - 5)$ using the method of Sec. 4.3. In this case, the roots are $x = \frac{2}{3}$ and $x = \frac{5}{4}$. Observe that $a_2 = 12$ is a multiple of each denominator (3 and 4) and that $a_0 = 10$ is a multiple of each numerator (2 and 5).

When there are irrational roots, if one root has the form $d + e\sqrt{f}$, another root will have the form $d - e\sqrt{f}$ (if the polynomial has rational coefficients). Similarly, when there are complex roots, if one root has the form $g + ih$, where i is the imaginary number (Sec. 3.1), then another root will have the form $g - ih$ (if the polynomial has real coefficients). We call these **conjugate** pairs.

Sometimes, sketching a graph of a polynomial can help you find the roots faster. (We will discuss graphs in Chapter 7.)

Keep in mind that there can be **double roots** or **triple roots**, for example (that is, where the roots are not all distinct; we call this **multiplicity**). For example, $x^2 - 6x + 9$ has a double root $(x - 3)(x - 3)$, meaning that both roots are the same: $x = 3$.

Example 1. Find all of the roots of $p(x) = x^3 + 23x^2 + 140x + 196$.
There will be 3 roots because $p(x)$ has degree 3 (since its highest power is x^3). Since all four coefficients are positive, if there are any real roots, they must be negative. Since $a_0 = 196$, if there are any integer roots, they must be factors of 196. Note that 196 factors as $2 \times 2 \times 7 \times 7 = 196$. Combining both ideas together, the possible integer roots include $-1, -2, -4, -7, -14, -28, -49, -98$, and -196. Let's try some and see if any of these happen to be roots:

- $p(-1) = (-1)^3 + 23(-1)^2 + 140(-1) + 196 = -1 + 23 - 140 + 196 = 78$.
- $p(-2) = (-2)^3 + 23(-2)^2 + 140(-2) + 196 = -8 + 92 - 280 + 196 = 0$.
- $p(-4) = (-4)^3 + 23(-4)^2 + 140(-4) + 196 = -64 + 368 - 560 + 196 = -60$.
- $p(-7) = (-7)^3 + 23(-7)^2 + 140(-7) + 196 = -343 + 1127 - 980 + 196 = 0$.
- $p(-14) = (-14)^3 + 23(-14)^2 + 140(-14) + 196 = -2744 + 4508 - 1960 + 196 = 0$.

We found all three roots already: $x = -2$, $x = -7$, and $x = -14$ are roots since $p(-2) = p(-7) = p(-14) = 0$.
Check: $(x + 2)(x + 7)(x + 14) = (x^2 + 9x + 14)(x + 14) = x^3 + 9x^2 + 14x + 14x^2 + 126x + 196 = x^3 + 23x^2 + 140x + 196$.

Example 2. Find all of the roots of $p(x) = 12x^3 + 65x^2 - 54x - 72$.

There will be 3 roots because $p(x)$ has degree 3 (since its highest power is x^3). Get started by checking a few easy values: $p(0) = -72, p(1) = 12 + 65 - 54 - 72 = -49$, and $p(-1) = -12 + 65 + 54 - 72 = 35$. Right away we know that at least one root lies between $x = -1$ and $x = 0$ because $p(-1)$ is positive and $p(0)$ is negative. This root has to be a fraction because there aren't any integers between $x = -1$ and $x = 0$.

Note that $a_3 = 12$ and $a_0 = -72$. If the fraction $\frac{N}{D}$ is a root of $p(x)$, $a_3 = 12$ will be evenly divisible by D and $a_0 = -72$ will be evenly divisible by N. Since we reasoned out that at least one fractional root lies between $x = -1$ and $x = 0$, this fraction could be $-\frac{1}{12}, -\frac{1}{6}, -\frac{1}{4}, -\frac{1}{3}, -\frac{1}{2}, -\frac{2}{3}$, or $-\frac{3}{4}$. (Why? The denominator D is a factor of 12, which could be 2, 3, 4, 6, or 12. The numerator N is a factor of 72, but must be less than D in order for $\frac{N}{D}$ to lie between $x = -1$ and $x = 0$. For example, when $N = 8$ and $D = 12$, the fraction is $\frac{N}{D} = \frac{8}{12} = \frac{2}{3}$.) Comparing $p(0) = -72$ with $p(-1) = 35$ suggests that the root might be closer to $x = -1$ than it is to $x = 0$, so let's start our search with $-\frac{3}{4}$ (and, if necessary, get less negative from there):

- $p\left(-\frac{3}{4}\right) = 12\left(-\frac{3}{4}\right)^3 + 65\left(-\frac{3}{4}\right)^2 - 54\left(-\frac{3}{4}\right) - 72 = 0$

We found one root already: $x = -\frac{3}{4}$ is a root since $p\left(-\frac{3}{4}\right) = 0$. Before we search for additional roots, we will divide $p(x)$ by $x - \left(-\frac{3}{4}\right)$, which is the same as dividing $p(x)$ by $x + \frac{3}{4}$, using synthetic division (Sec. 6.4). Recall from Sec. 64 that $c = -\frac{3}{4}$ for $x + \frac{3}{4}$.

$$
\begin{array}{r|rrrr}
-\frac{3}{4} & 12 & 65 & -54 & -72 \\
 & & -9 & -42 & -72 \\
\hline
 & 12 & 56 & -96 & 0
\end{array}
$$

The synthetic division above shows that $q(x) = \frac{p(x)}{x+\frac{3}{4}} = 12x^2 + 56x - 96$. Since $12, 56$, and 96 are each evenly divisible by 4, we may factor it out: $q(x) = 4(3x^2 + 14x - 24)$. Factor $3x^2 + 14x - 24$ using the method from Sec. 4.3 to get $q(x) = 4(x + 6)(3x - 4)$. The roots of $q(x)$ are $x = -6$ and $x = \frac{4}{3}$. We now know all three roots of $p(x)$. They are $x = -6$, $x = -\frac{3}{4}$, and $x = \frac{4}{3}$.

Checks: $p(-6) = 12(-6)^3 + 65(-6)^2 - 54(-6) - 72 = 0$,

$p\left(-\frac{3}{4}\right) = 12\left(-\frac{3}{4}\right)^3 + 65\left(-\frac{3}{4}\right)^2 - 54\left(-\frac{3}{4}\right) - 72 = 0$, and

$p\left(\frac{4}{3}\right) = 12\left(\frac{4}{3}\right)^3 + 65\left(\frac{4}{3}\right)^2 - 54\left(\frac{4}{3}\right) - 72 = 0$.

Observe that $4(x + 6)\left(x + \frac{3}{4}\right)(3x - 4) = (x + 6)(4x + 3)(3x - 4) =$

$(4x^2 + 27x + 18)(3x - 4) = 12x^3 + 81x^2 + 54x - 16x^2 - 108x - 72 =$

$12x^3 + 65x^2 - 54x - 72 = p(x)$.

Example 3. Find all of the roots of $p(x) = 12x^3 - 10x^2 + 12x - 10$.

There will be 3 roots because $p(x)$ has degree 3 (since its highest power is x^3). Since the odd terms ($12x^3$ and $12x$) are both positive while even terms ($-10x^2$ and -10) are both negative, if there are any real roots, they won't be negative. Since $p(0) = -10$ and $p(1) = 12 - 10 + 12 - 10 = 4$ have opposite signs, we know that at least one root lies between $x = 0$ and $x = 1$. This root must be a fraction. Note that $a_3 = 12$ and $a_0 = -10$. If the fraction $\frac{N}{D}$ is a root of $p(x)$, $a_3 = 12$ will be evenly divisible by D and $a_0 = -10$ will be evenly divisible by N. Since we reasoned out that at least one fractional root lies between $x = 0$ and $x = 1$, this fraction could be $\frac{1}{12}, \frac{1}{6}, \frac{1}{4}, \frac{1}{3}, \frac{1}{2}, \frac{5}{12}, \frac{2}{3}$, or $\frac{5}{6}$. (Why? The denominator D is a factor of 12, which could be 2, 3, 4, 6, or 12. The numerator N is a factor of 10, which could be 1, 2, 5, or 10, but must be less than D in order for $\frac{N}{D}$ to lie between $x = 0$ and $x = 1$. For example, when $N = 10$ and $D = 12$, the fraction is $\frac{N}{D} = \frac{10}{12} = \frac{5}{6}$.) Comparing $p(0) = -10$ with $p(1) = 4$ suggests that the root might be closer to $x = 1$ than it is to $x = 0$, so let's start our search with $\frac{5}{6}$:

- $p\left(\frac{5}{6}\right) = 12\left(\frac{5}{6}\right)^3 - 10\left(\frac{5}{6}\right)^2 + 12\left(\frac{5}{6}\right) - 10 = 0$

We found one root already: $x = \frac{5}{6}$ is a root since $p\left(\frac{5}{6}\right) = 0$. Before we search for additional roots, we will divide $p(x)$ by $x - \frac{5}{6}$ using synthetic division (Sec. 6.4). Recall from Sec. 6.4 that $c = \frac{5}{6}$ is positive for $x - \frac{5}{6}$.

$$\begin{array}{r|rrrr} \frac{5}{6} & 12 & -10 & 12 & -10 \\ & & 10 & 0 & 10 \\ \hline & 12 & 0 & 12 & 0 \end{array}$$

The synthetic division above shows that $q(x) = \frac{p(x)}{x-\frac{5}{6}} = 12x^2 + 0x + 12 = 12x^2 + 12$,

which factors as $q(x) = 12(x^2 + 1)$. As always, the roots of $q(x)$ are the values of x that make $q(x)$ zero. In this case, these are the values of x that make $x^2 + 1$ equal to zero. The roots of $q(x)$ are complex numbers (Chapter 3) because no real value squared plus one can be zero. We can solve the equation $x^2 + 1 = 0$ using the method from Sec. 3.10 to get $x^2 = -1$, for which the solutions are $x = \pm i$. It is easy to check that $x = \pm i$ satisfies $x^2 + 1 = 0$ because $i^2 = -1$. The roots of $q(x)$ are $x = \pm i$. We now know all three roots of $p(x)$. They are $x = \frac{5}{6}$, $x = -i$, and $x = i$.

Checks: $p\left(\frac{5}{6}\right) = 12\left(\frac{5}{6}\right)^3 - 10\left(\frac{5}{6}\right)^2 + 12\left(\frac{5}{6}\right) - 10 = 0$,

$p(-i) = 12(-i)^3 - 10(-i)^2 + 12(-i) - 10 = 12(-1)i^3 - 10(-1) - 12i - 10 = -12(-i) + 10 - 12i - 10 = 12i + 10 - 12i - 10 = 0$, and

Recall from Sec. 3.4 that $i^3 = -i$ such that $(-i)^3 = (-1)^3 i^3 = (-1)(-i) = i$. Here, we used the rule $(ab)^3 = a^3 b^3$ with $a = -1$ and $b = i$.

Observe that $12\left(x - \frac{5}{6}\right)(x - i)(x + i) = 2(6x - 5)(x^2 + 1) = 2(6x^3 - 5x^2 + 6x - 5) = 12x^3 - 10x^2 + 12x - 5 = p(x)$, where the 12 came from factoring $q(x)$.

Example 4. Find all of the roots of $p(x) = 6x^4 - 12x^2$.

There will be 4 roots because $p(x)$ has degree 4 (since its highest power is x^4). Since there is no constant term, $x = 0$ is a root. Actually, $x = 0$ is a double root because we can factor x^2 out of $p(x)$ to get $p(x) = x^2(6x^2 - 12)$. Since $(x - 0)(x - 0) = x^2$, this is how two of the four roots are $x = 0$. We can actually factor $6x^2$ out to get $p(x) = 6x^2(x^2 - 2)$. The other two roots are $x = \pm\sqrt{2}$.

Checks: $p(0) = 0$ and $p(\pm\sqrt{2}) = 6(\pm\sqrt{2})^4 - 12(\pm\sqrt{2})^2 = 6(2)^2 - 12(2) = 6(4) - 24 = 24 - 24 = 0$. Recall from Sec. 2.1 that $(\sqrt{2})^4 = (2^{1/2})^4 = 2^{4/2} = 2^2 = 4$.

6.7 Problems

Directions: Find all of the roots for each polynomial.

(1) $x^3 - 15x^2 + 71x - 105$

(2) $4x^3 + 88x^2 + 460x + 600$

(3) $x^3 + 3x^2 - 22x - 24$

(4) $x^4 + 6x^3 - 45x^2 - 122x + 360$

(5) $x^4 - 18x^3 + 116x^2 - 318x + 315$ (6) $x^4 - 8x^3 - 36x^2 + 288x$

(7) $30x^3 - 23x^2 - 7x + 6$ (8) $15x^3 - 41x^2 - 44x + 96$

(9) $12x^3 + 7x^2 - 77x - 72$

(10) $6x^3 - 30x^2 + 54x - 270$

(11) $3x^3 + 27x^2 - 36x - 324$

(12) $15x^5 - 10x^4 + 30x^3 - 20x^2$

7 GRAPHS

7.1 Graphing Functions and Equations

Recall from Chapter 1 that a function $f(x)$ has a single value of f for each value of x. In this chapter, we will graph functions as well as equations that aren't functions. The way that we graph functions and other equations is the same; the only distinction is that when graphing an equation for y that isn't a function (such as the equation for a circle), the graph may have multiple values of y for a single value of x.

Whether graphing a function or an equation that isn't a function, a good place to begin is to make a table of values, as shown in the examples. It may also help to identify key features, such as intercepts, roots (recall Sec. 6.7), minimum values, maximum values, or asymptotes (Sec. 7.8 discusses asymptotes for hyperbolas). The y-**intercept** is the point where the curve crosses the y-axis (meaning that x is zero), while the x-**intercept** is the point where the curve crosses the x-axis (meaning that y is zero).

Example 1. Graph $f(x) = 3 - x^2$.

Use the given equation to make a table. For example, $f(1) = 3 - 1^2 = 2$. Recall Sec. 1.3.

x	-5	-4	-3	-2	-1	0	1	2	3	4	5
y	-22	-13	-6	-1	2	3	2	-1	-6	-13	-22

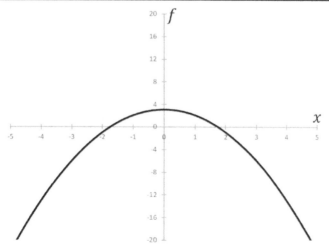

Observe that the f-intercept (since in this case, f appears on the vertical axis; this equation has x and f instead of the usual x and y) is $f(0) = 3 - 0^2 = 3$. To find the x-intercept, set f equal to zero to get $0 = 3 - x^2$. Solve this equation to find that there are two x-intercepts: $x = \pm\sqrt{3}$. On the graph, the x-intercepts lie at $\left(\pm\sqrt{3}, 0\right)$ and the f-intercept lies at $(0, 3)$. Like all functions, the graph of $f(x)$ passes the **vertical line test**: You can't draw a vertical line that intercepts two different points on the curve (because there is only one value of f for any given value of x). As a counterexample, see the graph in Example 2.

Example 2. Graph $x^2 + y^2 = 25$.
We could make a table again, but this graph is fairly easy to reason out. This equation resembles the Pythagorean theorem if we write it as $x^2 + y^2 = 5^2$. This means that $\sqrt{x^2 + y^2} = 5$ is the distance from the origin to any point on the curve, (x, y). What kind of curve is always 5 units from the origin? A circle with a radius equal to 5. Note that this equation does not describe a function because it doesn't pass the vertical line test. For most values of x on the circle, there are two corresponding values of y. For example, when $x = 3$, y can be 4 or -4 since the points $(3, 4)$ and $(3, -4)$ both lie on the circle. The y-intercepts lie at $(0, \pm 5)$ while the x-intercepts lie at $(\pm 5, 0)$.

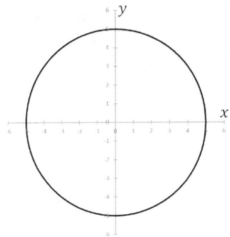

7.1 Problems

Directions: Sketch a graph for each equation. For each graph, use the vertical line test.

(1) $y = \dfrac{x^2}{2} - 3$

(2) $y = \dfrac{x^3}{8}$

(3) $y = x^4 - 2x^3 - 13x^2 + 14x + 24$

(4) $y = \sqrt{x}$

(5) $\dfrac{x^2}{9} + \dfrac{y^2}{16} = 1$

(6) $\dfrac{y^2}{16} - \dfrac{x^2}{9} = 1$

(7) $y = (x - 3)^2$

(8) $x = y^2$

(9) $y = x^5$

(10) $y = |x|$

(11) $xy = 1$

(12) $y = 2^x$

7.2 Straight Lines

An equation of the form $y = mx + b$ represents a straight line where m is the **slope** (which is a measure of the steepness of the line) and b is the y-**intercept**. Given the coordinates of any two points on the line, **slope** can be found from the rise over the run, which equals $m = \frac{y_2 - y_1}{x_2 - x_1}$. The y-**intercept** (b) is the value of y when x equals zero; this is the point where the line crosses the y-axis. The x-**intercept** is the value of x when y equals zero; this is the point where the line crosses the x-axis. To find the x-intercept from the equation for the line, set y equal to zero and solve for x. For the line below on the left, the y-intercept is $b = 2$, the x-intercept is -4, and the slope is $m = \frac{4-2}{4-0} = \frac{2}{4} = \frac{1}{2}$ using the indicated points $(0, 2)$ and $(4, 4)$. The rise is 2 (from 2 to 4) and the run is 4 (from 0 to 4).

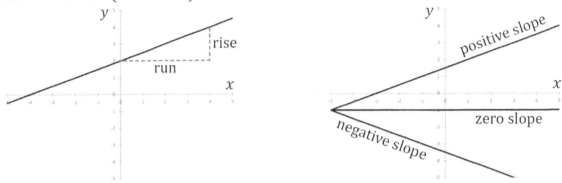

A **horizontal** line has zero slope and has the form $y = b$. A **vertical** line has infinite slope and has the form $x = c$. For example, the horizontal line above on the right is $y = -1$ and the vertical line coinciding with the y-axis is $x = 0$.

The equation $y = mx + b$ is the **slope-intercept** form of a straight line. An alternative is the **point-slope** form $y - y_1 = m(x - x_1)$ where (x_1, y_1) are the coordinates of any point on the line.

If two lines intersect, the coordinates of the point of intersection will satisfy both of the equations. One way to solve for the point of intersection is to set $mx + b$ equal for both lines (see Example 4) since they have the same value of y where they intersect. This allows you to solve for x.

The **origin** refers to the point $(0, 0)$.

Note: This book focuses on intermediate algebra skills. For students who need more basic graphing practice (like plotting points, graphing lines, finding the slope of lines shown in graphs, or writing the equations for lines shown in graphs), the author has a book dedicated to this: *Basic Linear Graphing Skills Practice Workbook*. The examples and problems in this chapter pick up where that book leaves off.

Example 1. Find the equation for a line passing through $(3, 1)$ and $(6, 13)$.
First use the given points to find the slope of the line.
$$m = \frac{y_2 - y_1}{x_2 - x_1} = \frac{13 - 1}{6 - 3} = \frac{12}{3} = 4$$
Plug $m = 4$ into the equation for a line: $y = 4x + b$. Plug either point into this to solve for the y-intercept. Plug in $x = 3$ and $y = 1$ for the point $(3, 1)$ to get $1 = 4(3) + b$. Simplify: $1 = 12 + b$ which leads to $-11 = b$. Plug this into $y = 4x + b$ to get $y = 4x - 11$. Use the point $(6, 13)$ to check the answer: $4(6) - 11 = 24 - 11 = 13$.

Example 2. A line with equation $y = \frac{x}{3} - 4$ passes through $(c, -1)$. What is c?
Plug $x = c$ and $y = -1$ for the given point into the equation: $-1 = \frac{c}{3} - 4$. Add 4 to both sides: $3 = \frac{c}{3}$. Multiply by 3 on both sides: $9 = c$.

Example 3. A line with a slope of 2 passes through $(2, 5)$ and $(7, d)$. What is d?
Use the slope equation: $m = \frac{y_2 - y_1}{x_2 - x_1} = \frac{d - 5}{7 - 2} = \frac{d - 5}{5}$. Plug in $m = 2$ to get $2 = \frac{d - 5}{5}$. Multiply by 5 on both sides: $10 = d - 5$. Add 5 to both sides: $15 = d$.

Example 4. Where do the lines $y = 7x + 10$ and $y = 3x - 6$ intersect?
Since both lines have the same value of y where they intersect, we may set the right-hand sides of these equations equal: $7x + 10 = 3x - 6$. Subtract 10 and subtract $3x$ from both sides: $4x = -16$. Divide by 4 on both sides: $x = -\frac{16}{4} = -4$. Plug $x = -4$ into either equation: $y = 7(-4) + 10 = -28 + 10 = -18$. The point of intersection is $(-4, -18)$. Check that this point satisfies the other equation: $3(-4) - 6 = -12 - 6 = -18$.

7.2 Problems

Directions: Find the equation for the line passing through each pair of points.

(1) $(2, 3)$ and $(5, 12)$.

(2) $(-2, -6)$ and $(1, 0)$.

(3) $(-1, 9)$ and $(1, -3)$.

(4) $(3, 5)$ and $(7, 7)$.

(5) $(0, 6)$ and $(-5, -9)$.

(6) $(-4, 11)$ and $(-1, -1)$.

Directions: Find the unknown coordinate.

(7) $y = 2x + 3$ passes through $(a, 15)$.

(8) $y = -x$ passes through $(-4, c)$.

(9) $y = 4x - 31$ passes through $(5, d)$.

(10) $y = \frac{x}{2} + 8$ passes through $(e, 12)$.

(11) $y = -\frac{x}{3} - 6$ passes through $(f, -9)$.

(12) $y = \frac{5x}{4}$ passes through $(-8, g)$.

Directions: A line with the given slope passes through the given points. Find the unknown.

(13) $(5, 3)$, $(a, 12)$, slope $= 3$

(14) $(-2, 4)$, $(5, c)$, slope $= -1$

(15) $(d, -10)$, $(3, 15)$, slope $= 5$

(16) $(1, e)$, $(9, 4)$, slope $= \frac{1}{2}$

(17) $(-7, 11)$, $(5, f)$, slope $= 0$

(18) $(0, 8)$, $(g, -16)$, slope $= -6$

Directions: Find the x-intercept and y-intercept for each line.

(19) $y = 6x + 8$

(20) $y = -\frac{x}{4}$

(21) $y = 12 - 3x$

(22) $y = 6x - 2$

(23) $y = -4x - 24$

(24) $y - 8 = 4(x - 3)$

Directions: Find the x-intercept and y-intercept for the line passing through the points.

(25) $(2, 12)$ and $(8, 0)$.

(26) $(3, 6)$ and $(6, 3)$.

(27) $(-4, 8)$ and $(8, -4)$.

(28) $(0, 24)$ and $(6, 9)$.

(29) $(-6, -3)$ and $(2, 5)$.

(30) $(-3, -9)$ and $(1, -6)$.

Directions: A line with the given intercept passes through the given points. Find the unknown.

(31) $(3, 2)$, $(9, a)$, y-intercept $= 6$

(32) $(2, c)$, $(8, 11)$, y-intercept $= 3$

(33) $(d, 18)$, $(-4, 3)$, y-intercept $= -3$

(34) $(6, 3)$, $(e, 15)$, x-intercept $= 9$

(35) $(1, f)$, $(3, 5)$, x-intercept $= 4$

(36) $(g, 10)$, $(-3, 4)$, x-intercept $= -5$

Pay close attention: Exercises 34-36 give the x-intercept (NOT the y-intercept).

Directions: Find the equation of the line that has the given intercepts.

(37) x-intercept $= 4$, y-intercept $= 4$ (38) x-intercept $= -2$, y-intercept $= 4$

(39) x-intercept $= 3$, y-intercept $= -9$ (40) x-intercept $= 12$, y-intercept $= 3$

(41) x-intercept $= -6$, y-intercept $= 2$ (42) x-intercept $= -10$, y-intercept $= -5$

Directions: Find the point of intersection for each pair of lines.

(43) $y = 4x + 5$ and $y = 7x - 16$ (44) $y = 2x - 9$ and $y = 6x - 1$

(45) $y = \frac{x}{3} - 8$ and $y = \frac{x}{2} + 3$ (46) $y = 5x - 15$ and $y = 9 - x$

(47) $y = -3x + 8$ and $x = -3$ (48) $y - 3 = 4(x - 7)$ and $y = 9x + 20$

7.3 Parallel and Perpendicular Lines

If two lines are **parallel**, they have the same slope. For example, $y = 7x - 4$ is parallel to $y = 7x + 3$ because both lines have slope $m = 7$. If two lines are **perpendicular**, their slopes satisfy the equation $m_1 m_2 = -1$, which is equivalent to $m_2 = -\frac{1}{m_1}$. For example, $y = \frac{2}{3}x + 5$ is perpendicular to $y = -\frac{3}{2}x + 4$ because $m_1 = \frac{2}{3}$, $m_2 = -\frac{3}{2}$, and $-\frac{1}{m_1} = -\frac{1}{2/3} = -1 \div \frac{2}{3} = -1 \times \frac{3}{2} = -\frac{3}{2} = m_2$. The main idea is that $m_1 m_2 = \frac{2}{3}\left(-\frac{3}{2}\right) = -1$. Another way to put it is that m_1 and m_2 are **negative reciprocals** of one another.

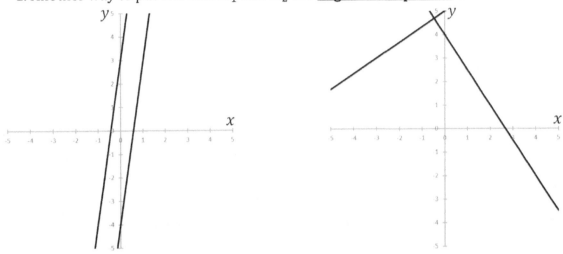

For a parallel line, the slope is the same: $y = 2x + b$. Plug in the given coordinates to solve for the y-intercept: $7 = 2(4) + b$. Simplify: $7 = 8 + b$. Subtract 8 from both sides: $-1 = b$. The equation for the line is $y = 2x - 1$.

Example 2. Find the line that is perpendicular to $y = 2x + 5$ and passes through $(4, 7)$. Take the negative reciprocal of the slope to get the new slope: $m = -\frac{1}{2}$. The equation for the line is $y = -\frac{x}{2} + b$. Plug in the given coordinates to solve for the y-intercept: $7 = -\frac{4}{2} + b$. Simplify: $7 = -2 + b$. Add 2 to both sides: $9 = b$. The equation for the line is $y = -\frac{x}{2} + 9$.

7.3 Problems

Directions: Find the equation of the line that satisfies the criteria.

(1) parallel to $y = 6x - 3$
and passes through $(5, 9)$

(2) perpendicular to $y = x + 2$
with a y-intercept of 4

(3) perpendicular to $y = 3x + 7$
and passes through the origin

(4) parallel to $y = -2x$
with an x-intercept of 9

(5) parallel to $y = 9x + 5$
with an x-intercept of 6

(6) perpendicular to $y = -\frac{3x}{4} + 1$
and passes through $(12, -24)$

(7) perpendicular to $y = \frac{x}{5} - 8$
and passes through $(-3, 20)$

(8) perpendicular to $y = 5$
and passes through $(-3, 7)$

(9) perpendicular to $y = \frac{2x}{5} + 4$

and passes through $(2, 9)$

(10) perpendicular to $y = -\frac{x}{8}$

and passes through $(-6, 4)$

(11) parallel to $y = 9 - 6x$

with a y-intercept of -12

(12) perpendicular to $y = -7x$

with an x-intercept of 14

(13) perpendicular to $y = \frac{3x}{5} + \frac{4}{5}$

and passes through $\left(-\frac{1}{5}, \frac{5}{3}\right)$

(14) parallel to $y = -\frac{x}{6} + 3$

with an x-intercept of $\frac{2}{3}$

(15) perpendicular to $y = 4x - 12$

with an x-intercept of -4

(16) perpendicular to $y = 0.4x$

and passes through $(0.8, -1.2)$

(17) parallel to $x = -3$

and passes through $(-1, -5)$

(18) perpendicular to $y = x\sqrt{3}$

with an x-intercept of $\sqrt{3}$

7.4 Linear Relationships

If two quantities have a linear relationship, they satisfy the equation $y = mx + b$. We may apply what we know about the equation for a straight line to solve the problem. For example, any two points on the line can be used to find the slope: $m = \frac{y_2 - y_1}{x_2 - x_1}$. The equation for a straight line still has the same form even if different variables are used. For example, if a has a linear relationship with t, we may write $a = mt + b$ and find the slope using $m = \frac{a_2 - a_1}{t_2 - t_1}$. Note that a function may also be linear. For example, $f(x) = mx + b$ is a linear function of x (recall Chapter 1). For this function, the slope is $m = \frac{f(x_2) - f(x_1)}{x_2 - x_1}$ where $f(x_2)$ and $f(x_1)$ mean to evaluate f at the respective values of x (Sec. 1.3). They do NOT mean to multiply f by a value of x.

Example 1. A car's speed increases linearly with time. At 3 seconds, the car's speed is 50 m/s. At 7 seconds, the car's speed is 74 m/s. What is the car's speed at 10 seconds? What was the car's speed at 0 seconds?

The speed v and time t have a linear relationship: $v = mt + b$. (It is common to use the letter v for speed instead of s. The reason has to do with the related term "velocity.") Use the given numbers to find the slope: $m = \frac{v_2 - v_1}{t_2 - t_1} = \frac{74 - 50}{7 - 3} = \frac{24}{4} = 6$. (In this case, the slope represents the car's acceleration, though it isn't necessary to know anything about acceleration or even speed to solve this problem.) Plug in one of the coordinate pairs to solve for the y-intercept (which in this case is really the v-intercept). We will use the point $(3, 50)$. Plug $t = 3$ and $v = 50$ along with $m = 6$ into $v = mt + b$ to get $50 = 6(3) + b$. Simplify: $50 = 18 + b$. Subtract 18 from both sides: $32 = b$ is the v-intercept. The equation is $v = 6t + 32$. Use this equation to answer the questions.

- At 10 seconds, $v = 6(10) + 32 = 60 + 32 = 92$ m/s is the car's speed.
- At 0 seconds, $v = 6(0) + 32 = 32$ m/s is the car's speed. (In this problem, the v-intercept represents the initial speed of the car.)

Example 2. A scientist determines that pressure is a linear function of temperature, $P(T)$, for some (non-ideal) gas. The scientist records the following data: $P(40 \text{ K}) = 300$ Pa and $P(60 \text{ K}) = 400$ Pa. Determine the pressure when the temperature is 90 K and 120 K. Note: Pa stands for Pascals and K stands for degrees Kelvin.

Since the relationship is linear, we may use the equation $P = mT + b$. Use the data to find the slope: $m = \frac{P_2 - P_1}{T_2 - T_1} = \frac{400 - 300}{60 - 40} = \frac{100}{20} = 5$. Plug one of the data points into the equation to determine the y-intercept (which in this case is really the P-intercept). We will use the point $(40, 300)$. Plug $T = 40$ and $P = 300$ along with $m = 5$ into $P = mT + b$ to get $300 = 5(40) + b$. Simplify: $300 = 200 + b$. Subtract 200 from both sides: $100 = b$ is the P-intercept. The equation is $P(T) = 5T + 100$. Use this equation to answer the questions.

- $P(90) = 5(90) + 100 = 450 + 100 = 550$ Pa.
- $P(120) = 5(120) + 100 = 600 + 100 = 700$ Pa.

7.4 Problems

Directions: Write an equation for each problem and use it to answer the questions.

(1) A train's position varies linearly with time. At 12 seconds, the train is 450 feet from the station. At 20 seconds, the train is 650 feet from the station. How far will the train be from the station at 30 seconds? Where was the train at 0 seconds? How fast is the train moving?

(2) A vending machine's daily profits depend linearly on the number of cans sold. When 25 cans are sold in one day, the profit is $45. When 40 cans are sold in one day, the profit is $90. What will the profit be if 60 cans are sold in one day? How many cans does it take to break even? What would the loss be if 8 cans were sold in one day?

(3) The sum of the interior angles of a polygon depends linearly on the number of sides. For a polygon with 5 sides, the interior angles add up to 540°. For a polygon with 7 sides, the interior angles add up to 900°. What is the sum of the interior angles for a decagon (which has 10 sides) and for a triangle (which has 3 sides)?

(4) A box of identical balls weighs 600 ounces when it contains 40 balls and weighs 750 ounces when it contains 90 balls. How much will the box weigh if it contains 200 balls? How much does the empty box weigh? How much does one ball weigh? If the box weighs 1500 ounces when it is full, how many balls can the box hold?

(5) Water drains from a tub at a linear rate. The tub contains 48 gallons of water at 10 minutes and contains 45 gallons of water at 16 minutes. How much water will the tub contain at 32 minutes? How much water was in the tub at zero minutes? When will the tub be empty?

(6) A store's profit per item depends linearly on the selling price. If an item sells for 60 cents, the store makes a profit of 15 cents. If an item sells for 90 cents, the store makes a profit of 30 cents. How much profit will the store earn if it sells an item for $1.80? How much does an item sell for when the store makes a profit of 50 cents?

(7) A car slows down at a linear rate. The car's speed is 40 m/s at 9 seconds and 30 m/s at 12 seconds. What will the car's speed be at 18 seconds? When will the car come to rest? What was the car's speed at zero seconds?

(8) In the photoelectric effect, the maximum kinetic energy of the photoelectrons is a linear function of the frequency: $K(f)$. Given that $K(0) = -2$ eV and $K(300 \text{ THz}) = 0$, find the maximum kinetic energy of the photoelectrons when the frequency is 800 THz and find the frequency for which the maximum kinetic energy of the photoelectrons is 6 eV. Note: THz stands for terahertz and eV stands for electron Volt.

(9) The mass of fuel in a rocket is a linear function of the burn time: $m(t)$. Given that $m(7 \text{ s}) = 500$ kg and $m(12 \text{ s}) = 250$ kg, when will the rocket run out of fuel? How much fuel did the rocket have at zero seconds? Note: s is a second and kg is a kilogram.

(10) The value of stock A is increasing linearly while the value of stock B is decreasing linearly. Stock A's value was $55 on day 22 and $67 on day 28. Stock B's value was $141 on day 14 and $125 on day 30. At this rate, on which day will the value of stocks A and B be the same?

7.5 Circles

The equation $(x - a)^2 + (y - b)^2 = R^2$ represents a **circle** with **radius** R and with its **center** at (a, b). Every point that lies on the edge of the circle is equidistant from the **center**. The **radius** is the distance from the center to any point on the edge of the circle. The **diameter** is twice the radius: $D = 2R$. The **circumference** is the total distance along the complete arc of the circle and is related to the diameter and radius by $C = \pi D = 2\pi R$, where the lowercase Greek letter pi approximately equals $\pi \approx 3.14159265$. The **area** of the circle is $A = \pi R^2$. In the diagram below on the left, the circle is centered at $(3, 2)$ and has radius $R = 5$, diameter $D = 10$, and equation $(x - 3)^2 + (y - 2)^2 = 25$. Note that $R^2 = 5^2 = 25$, $a = 3$, and $b = 2$.

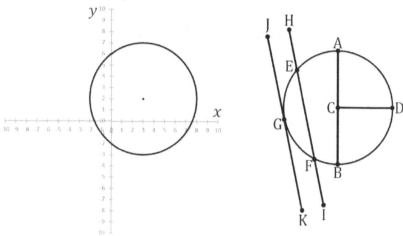

To solve for the points where two circles intersect or where a line intersects a circle, solve the (non-linear) system of two equations for the values of x and y that satisfy both of the equations. See Example 3.

Example 1. What is the radius of $(x - 7)^2 + (y - 5)^2 = 36$? Where is its center?
Compare $(x - 7)^2 + (y - 5)^2 = 36$ with $(x - a)^2 + (y - b)^2 = R^2$ to see that $a = 7$, $b = 5$, and $R^2 = 36$. The radius is $R = \sqrt{36} = 6$ and the center is at $(7, 5)$.

Example 2. What is the radius of $3x^2 + 3y^2 - 6x - 12y = 6$? Where is its center?
Rewrite the equation in standard form. Divide by 3 on both sides: $x^2 + y^2 - 2x - 4y = 2$. Use the technique from Sec. 4.4 called "completing the square." We need to add 1 to $x^2 - 2x$ because $(x - 1)^2 = x^2 - 2x + 1$ and we need to add 4 to $y^2 - 4y$ because $(y - 2)^2 = y^2 - 4y + 4$. Therefore, we add $1 + 4 = 5$ to both sides of the equation for the circle to get $x^2 - 2x + 1 + y^2 - 4y + 4 = 7$, which equates to $(x - 1)^2 + (y - 2)^2 = 7$. You can use the FOIL method to verify that $(x - 1)^2 + (y - 2)^2 = 7$ agrees with $x^2 + y^2 - 2x - 4y = 2$. Compare $(x - 1)^2 + (y - 2)^2 = 7$ with $(x - a)^2 + (y - b)^2 = R^2$ to see that $a = 1$, $b = 2$, and $R^2 = 7$. The radius is $R = \sqrt{7}$ and the center is at $(1, 2)$.

Example 3. Find the points where $y = 3x + 1$ intersects $x^2 + y^2 = 4$.
Square both sides of the equation for the line: $y^2 = 9x^2 + 6x + 1$. Replace y^2 with $9x^2 + 6x + 1$ in the equation for the circle: $x^2 + 9x^2 + 6x + 1 = 4$. Subtract 4 from both sides and combine like terms: $10x^2 + 6x - 3 = 0$. Use the quadratic formula.

$$x = \frac{-6 \pm \sqrt{6^2 - 4(10)(-3)}}{2(10)} = \frac{-6 \pm \sqrt{36 + 120}}{20} = \frac{-6 \pm \sqrt{156}}{20}$$

$$x = \frac{-6 \pm \sqrt{4(39)}}{20} = \frac{-6 \pm 2\sqrt{39}}{20} = \frac{-3 \pm \sqrt{39}}{10}$$

Plug this into the equation for the line to solve for y.

$$y = 3\left(\frac{-3 \pm \sqrt{39}}{10}\right) + 1 = \frac{-9 \pm 3\sqrt{39}}{10} + \frac{10}{10} = \frac{1 \pm 3\sqrt{39}}{10}$$

Calculator check: $x \approx 0.3245$ and $y \approx 1.973$ or $x \approx -0.9245$ and $y \approx -1.773$.

$$3x + 1 \approx 3(0.3245) + 1 \approx 1.974 \approx y$$
$$x^2 + y^2 \approx 0.3245^2 + 1.973^2 \approx 3.998 \approx 4$$
$$3x + 1 \approx 3(-0.9245) + 1 \approx -1.774 \approx y$$
$$x^2 + y^2 \approx (-0.9245)^2 + (-1.773)^2 \approx 3.998 \approx 4$$

7.5 Problems

Directions: For each circle, determine the radius and the coordinates of its center.

(1) $(x - 3)^2 + (y - 8)^2 = 49$

(2) $x^2 + y^2 = 5$

(3) $(x + 9)^2 + (y - 4)^2 = 8$

(4) $x^2 + 8x + y^2 = 9$

(5) $x^2 - 16x + y^2 + 10y = 32$

(6) $2x^2 + 2y^2 = 54 + 24y - 36x$

Directions: Determine the points of intersection for each pair of equations.

(7) $x^2 + y^2 = 37$ and $y = 2x + 4$

(8) $(x - 1)^2 + y^2 = 4$ and $(x + 1)^2 + y^2 = 9$

(9) $x^2 + (y - 3)^2 = 45$ and $y = x$

(10) $x^2 + (y + 2)^2 = 25$ and $y = -x - 9$

7.6 Ellipses

The equation $\frac{x^2}{a^2} + \frac{y^2}{b^2} = 1$ represents an **ellipse** centered about the origin with a width of $2a$ and a height of $2b$. If $a > b$, then a is the **semi-major axis** (such that $2a$ is the major axis) and b is the **semi-minor axis** (such that $2b$ is the minor axis). If $a < b$, then these are reversed. An ellipse has two **foci** (which is the plural of **focus**), such that the sum of the distances from the two foci to any point on the ellipse is the same. If $a > b$, the foci are located at $(\pm c, 0)$, where $c = \sqrt{a^2 - b^2}$. If $a < b$, they are located at $(0, \pm c)$, where $c = \sqrt{b^2 - a^2}$. The **eccentricity** equals $e = \frac{c}{a}$ if $a > b$ or $e = \frac{c}{b}$ if $a < b$. The points at the ends of the major axis are called **vertices** (the plural of **vertex**). The **area** of an ellipse is $A = \pi ab$. In the diagram below, the equation of the ellipse is $\frac{x^2}{16} + \frac{y^2}{9} = 1$, the semi-major axis is BC = CE = a = 4 (such that $a^2 = 4^2 = 16$), the semi-minor axis is AC = CD = b = 3 (such that $b^2 = 3^2 = 9$), points B and E are vertices, points F and G are foci, point C lies at the center, CF = CG = $c = \sqrt{a^2 - b^2} = \sqrt{4^2 - 3^2} = \sqrt{7}$, the foci lie at $(\pm\sqrt{7}, 0)$, and FP + PG is the same for any point P that lies on the ellipse.

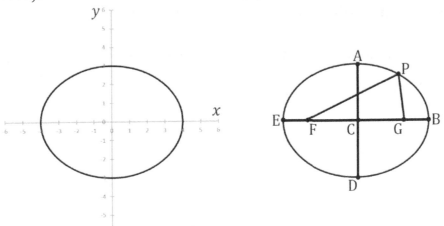

Example 1. Where are the foci of $\frac{x^2}{9} + \frac{y^2}{4} = 1$?

Compare $\frac{x^2}{9} + \frac{y^2}{4} = 1$ with $\frac{x^2}{a^2} + \frac{y^2}{b^2} = 1$ to see that $a = 3$ and $b = 2$. Use the formula $c = \sqrt{a^2 - b^2} = \sqrt{3^2 - 2^2} = \sqrt{9 - 4} = \sqrt{5}$. The foci are located at $(\pm\sqrt{5}, 0)$.

Example 2. Where are the foci of $\frac{x^2}{4} + \frac{y^2}{9} = 1$?

The only difference from this equation and the equation in Example 1 is that the roles of a and b are swapped. This time, the semimajor axis lies along y instead of x (since $a < b$). The foci are located at $\left(0, \pm\sqrt{5}\right)$. Compare this to the previous example.

7.6 Problems

Directions: Determine where the foci of each ellipse are located.

(1) $\frac{x^2}{25} + \frac{y^2}{16} = 1$

(2) $\frac{x^2}{3} + \frac{y^2}{4} = 1$

(3) $\frac{x^2}{9} + y^2 = 1$

(4) $\frac{x^2}{8} + \frac{y^2}{4} = 1$

(5) $10x^2 + 6y^2 = 60$

(6) $\frac{x^2}{169} + \frac{y^2}{144} = 1$

(7) $\frac{x^2}{225} + \frac{y^2}{289} = 1$

(8) $\frac{(x-3)^2}{25} + \frac{(y-2)^2}{9} = 1$

7.7 Parabolas

The equation $y = a(x - b)^2 + c$ represents a **parabola** with a vertical axis of symmetry about the line $x = b$. If $a > 0$, the parabola opens upward and its **vertex** is a minimum. If $a < 0$, the parabola opens downward and its **vertex** is a maximum. The **focus** lies at $\left(b, c + \frac{1}{4a}\right)$ and the **directrix** is the horizontal line $y = c - \frac{1}{4a}$. Any point on the parabola is equidistant from the focus and directrix. The equation $x = d(y - e)^2 + f$ represents a parabola with a horizontal axis of symmetry. A parabola that has the form $y = ux^2 + vx + w$ can be put into the standard form $y = a(x - b)^2 + c$ by completing the square (see Example 3). In the diagram below, the equation of the parabola is $y = \frac{1}{4}(x - 3)^2 + 2$, $a = \frac{1}{4}$, $b = 3$, $c = 2$, the parabola opens upward (since $a = \frac{1}{4} > 0$), the vertex at point B is a minimum, the line of symmetry is the vertical line $x = 3$, the focus (point A) lies at $(3, 3)$, the directrix is the horizontal line $y = 1$, and AP = CP (a similar relation holds for any point P on the parabola using the vertical distance to the directrix).

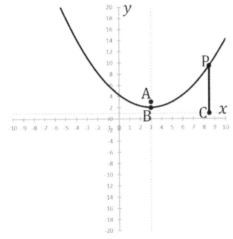

Example 1. Find the focus, directrix, and line of symmetry for $y = 2(x - 5)^2 + 7$. Compare $y = 2(x - 5)^2 + 7$ with $y = a(x - b)^2 + c$ to see that $a = 2$, $b = 5$, and $c = 7$. The line of symmetry is the vertical line $x = 5$. Since $c + \frac{1}{4a} = 7 + \frac{1}{4(2)} = \frac{56}{8} + \frac{1}{8} = \frac{57}{8}$ and $c - \frac{1}{4a} = 7 - \frac{1}{4(2)} = \frac{56}{8} - \frac{1}{8} = \frac{55}{8}$, the focus lies at $\left(5, \frac{57}{8}\right)$ and the directrix is the horizontal line $y = \frac{55}{8}$.

Example 2. Find the focus, directrix, and line of symmetry for $x = 2(y - 5)^2 + 7$. The only difference from this equation and the equation in Example 1 is that the roles of x and y are swapped. This time, the line of symmetry is the horizontal line $y = 5$, the focus lies at $\left(\frac{57}{8}, 5\right)$, and the directrix is the vertical line $x = \frac{55}{8}$. Compare this to the previous example.

Example 3. Find the focus, directrix, and line of symmetry for $y = 3x^2 + 12x - 2$. This equation is not in standard form. To put this equation in standard form, we must **complete the square** using the method from Sec. 4.4, but with an important difference. Since the standard form for a parabola, $y = a(x - b)^2 + c$, includes a constant a out front, we first need to **factor** out the 3 in the given equation: $y = 3\left(x^2 + 4x - \frac{2}{3}\right)$. Now **complete the square** inside of the parentheses. Compare with $(x + 2)^2 = x^2 + 4x + 4$. Rewrite $-\frac{2}{3}$ as $4 - 4 - \frac{2}{3}$, which equals $4 - \frac{12}{3} - \frac{2}{3} = 4 - \frac{14}{3}$.

$$x^2 + 4x - \frac{2}{3} = x^2 + 4x + 4 - \frac{14}{3} = (x + 2)^2 - \frac{14}{3}$$

Substitute the above expression into $y = 3\left(x^2 + 4x - \frac{2}{3}\right)$ to get the following:

$$y = 3\left(x^2 + 4x - \frac{2}{3}\right) = 3\left[(x + 2)^2 - \frac{14}{3}\right]$$

Now distribute the 3:

$$y = 3(x + 2)^2 - 14$$

Compare $y = 3(x + 2)^2 - 14$ with $y = a(x - b)^2 + c$ to see that $a = 3$, $b = -2$, and $c = -14$. Note the signs of b and c. The line of symmetry is the vertical line $x = -2$. Since $c + \frac{1}{4a} = -14 + \frac{1}{4(3)} = -\frac{168}{12} + \frac{1}{12} = -\frac{167}{12}$ and $c - \frac{1}{4a} = -10 - \frac{1}{4(3)} = -\frac{168}{12} - \frac{1}{12}$ $= -\frac{169}{12}$, the focus lies at $\left(-2, -\frac{167}{12}\right)$ and the directrix is the horizontal line $y = -\frac{169}{12}$. Check: $y = 3(x + 2)^2 - 14 = 3(x^2 + 4x + 4) - 14 = 3x^2 + 12x + 12 - 14 = 3x^2 + 12x - 2$.

Example 4. Find the points where $y = -2x + 4$ intersects $y = x^2 - 3x + 2$.

Since each equation equals y, the expressions on the right-hand side must be equal: $-2x + 4 = x^2 - 3x + 2$. Subtract 4 from both sides and add $2x$: $0 = x^2 - x - 2$. This quadratic equation can be factored using the method from Sec. 4.3: $0 = (x + 1)(x - 2)$. There are two possible answers for x. Either $x = -1$ or $x = 2$. Plug $x = -1$ into the equation for the line to get $y = -2(-1) + 4 = 2 + 4 = 6$ and plug $x = 2$ into the same equation to get $y = -2(2) + 4 = -4 + 4 = 0$. This shows that $x = -1$ and $y = 6$ is one solution and $x = 2$ and $y = 0$ is another solution. The two points of intersection are $(-1, 6)$ and $(2, 0)$.

7.7 Problems

Directions: For each parabola, find the focus, directrix, and line of symmetry.

(1) $y = 3(x - 2)^2 - 1$

(2) $x = \frac{(y+6)^2}{4} + 8$

(3) $y = -\frac{x^2}{3}$

(4) $x = 7y^2 - \frac{3}{4}$

(5) $y = 6x^2 + 24x + 5$

(6) $y = 5x^2 - 30x - 9$

Directions: Determine the points of intersection for each pair of equations.

(7) $y = x + 6$ and $y = x^2$

(8) $y = 3x + 5$ and $y = 2x^2 + 8x - 7$

(9) $y = 2x + 18$ and $x = 3y^2 + 6y - 7$

(10) $y = 2x^2 - 11x + 13$ and $y = -x^2 + 8x - 7$

(11) $x^2 + y^2 = 9$ and $y = \dfrac{x^2}{8}$

(12) $y = 4x^2$ and $x = 32y^2$

7.8 Hyperbolas

The equation $\frac{x^2}{a^2} - \frac{y^2}{b^2} = 1$ represents a **hyperbola** centered about the origin with its foci and vertices lying on the x-axis. This hyperbola has two **branches**: one for $x \geq a$ and one for $x \leq -a$. Its **vertices** are the two points closest to the y-axis: $(\pm a, 0)$. The difference (in absolute values) between the distances from the two **foci** to any point on the hyperbola is the same. The **foci** are located at $(\pm c, 0)$, where $c = \sqrt{a^2 + b^2}$. The hyperbola has two **asymptotes**, which are two lines that the branches of the hyperbola approach without ever quite reaching. The asymptotes are $y = \pm \frac{b}{a}x$. The special case $a = b$ is referred to as a **rectangular** hyperbola. For a rectangular hyperbola, $c = a\sqrt{2}$, such that the foci lie at $(\pm a\sqrt{2}, 0)$. The equation $y = \frac{1}{x}$ is a rectangular hyperbola with $a = b = \sqrt{2}$ that is rotated by $45°$; it has the x- and y-axes as asymptotes, vertices at $(1, 1)$ and $(-1, -1)$ such that the distance from the origin to a vertex equals $a = b = \sqrt{2}$, and foci at $(\sqrt{2}, \sqrt{2})$ and $(-\sqrt{2}, -\sqrt{2})$ such that the distance from the origin to a focus equals $c = a\sqrt{2} = \sqrt{2}\sqrt{2} = 2$ (consistent with $\sqrt{\left(\sqrt{2}\right)^2 + \left(\sqrt{2}\right)^2} = \sqrt{2 + 2} = 2$).

The equation $\frac{y^2}{a^2} - \frac{x^2}{b^2} = 1$ has the roles of x and y reversed compared to $\frac{x^2}{a^2} - \frac{y^2}{b^2} = 1$.

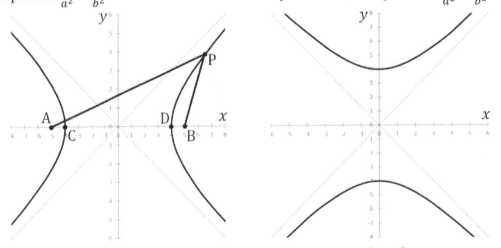

In the left diagram above, the equation of the hyperbola is $\frac{x^2}{16} - \frac{y^2}{9} = 1$, $a = 4$, $b = 3$, $c = \sqrt{a^2 + b^2} = \sqrt{4^2 + 3^2} = \sqrt{16 + 9} = \sqrt{25} = 5$, vertices C and D lie at $(\pm 4, 0)$, foci A and B lie at $(\pm 5, 0)$, $|AP - BP|$ is the same for any point P that lies on either branch of the hyperbola, and $y = \pm\frac{3x}{4}$ are asymptotes. In the right graph on the previous page, the equation of the hyperbola is $\frac{y^2}{16} - \frac{x^2}{9} = 1$ (the roles of x and y are reversed).

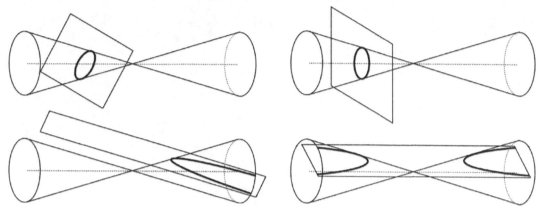

The ellipse, circle, parabola, and hyperbola are called **conic sections** because they can be formed by the intersection of a plane and a double cone, as illustrated above. The general equation $ax^2 + by^2 + cx + dy + e = 0$ can describe any of the conic sections. For example, if either a or b is zero (but not both), this equation describes a parabola. If $a = b$, it describes a circle. An ellipse can result when $ab > 0$ and a hyperbola can result when $ab < 0$. (The equation can also produce a line, a pair of lines, or a point.)

Example 1. Find the foci and asymptotes of $\frac{x^2}{5} - \frac{y^2}{4} = 1$.

Compare $\frac{x^2}{5} - \frac{y^2}{4} = 1$ with $\frac{x^2}{a^2} - \frac{y^2}{b^2} = 1$ to see that $a = \sqrt{5}$ and $b = 2$. Use the formula $c = \sqrt{a^2 + b^2} = \sqrt{\left(\sqrt{5}\right)^2 + 2^2} = \sqrt{5 + 4} = \sqrt{9} = 3$. The foci are located at $(\pm 3, 0)$. The asymptotes are $y = \pm\frac{b}{a}x = \pm\frac{2x}{\sqrt{5}} = \pm\frac{2x\sqrt{5}}{\sqrt{5}\sqrt{5}} = \pm\frac{2x\sqrt{5}}{5}$. In the last step, we rationalized the denominator (Sec. 2.4).

Example 2. Find the foci and asymptotes of $\frac{y^2}{5} - \frac{x^2}{4} = 1$.

The only difference from this equation and the equation in Example 1 is that the roles of x and y are swapped. The foci are located at $(0, \pm 3)$ and the asymptotes are $x = \pm\frac{2y\sqrt{5}}{5}$, which are equivalent to $y = \pm\frac{5x}{2\sqrt{5}}$, which simplifies to $y = \pm\frac{x\sqrt{5}}{2}$ since $\frac{5}{\sqrt{5}} = \sqrt{5}$.

7.8 Problems

Directions: For each hyperbola, find the foci and asymptotes.

(1) $\dfrac{x^2}{36} - \dfrac{y^2}{64} = 1$

(2) $\dfrac{x^2}{12} - \dfrac{y^2}{4} = 1$

(3) $\dfrac{y^2}{9} - \dfrac{x^2}{18} = 1$

(4) $\dfrac{x^2}{3} - y^2 = 1$

(5) $\dfrac{x^2}{144} - \dfrac{y^2}{25} = 1$

(6) $\dfrac{y^2}{64} - \dfrac{x^2}{225} = 1$

(7) $4x^2 - 3y^2 = 12$

(8) $y = -\dfrac{1}{x}$

7.9 Graphs of Even and Odd Functions

Recall the distinction between even and odd functions from Sec. 1.8. In this section, we will explore how these functions differ visually. If a function is **even**, the graph is unchanged if it is reflected about the y-axis, meaning that it is symmetric about the y-axis, like the graph of $y = x^2$ (left figure). If a function is **odd**, the graph is unchanged if it is rotated through $180°$ about the origin, like the graph of $y = x^3$ (right figure).

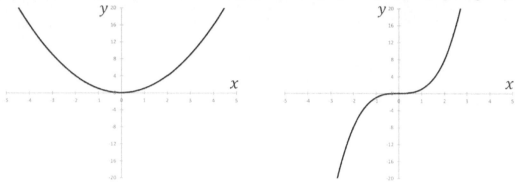

Example 1. $y(x) = x^2$ is an even function because $y(-x) = (-x)^2 = x^2 = y(x)$. This graph is symmetric about the y-axis (left figure).

Example 2. $y(x) = x^3$ is an odd function because $y(-x) = (-x)^3 = -x^3 = -y(x)$. If this graph is rotated $180°$ about the origin, the graph remains unchanged (right figure).

7.9 Problems

Directions: Sketch a graph for each equation. Is the function even, odd, or neither?

(1) $y(x) = x^4$

(2) $y(x) = x^5$

(3) $y(x) = x^4 - 3x^2 + 8$

(4) $y(x) = x^3 - 6x$

(5) $y(x) = \frac{1}{x}$

(6) $y(x) = \frac{1}{x^2}$

(7) $y(x) = |x|$

(8) $y(x) = x^2 - 4x + 3$

8 INEQUALITIES

8.1 Dividing Both Sides by Negative Coefficients

When **multiplying** or **dividing** both sides of an inequality by a **negative number**, the direction of the inequality changes. For example, if we divide by -2 on both sides of $-2x < 8$, we get $x > -4$. Why? One way to see the reason for this is to consider the simple inequality $2 < 3$, which is obviously true. If we multiply both sides by -1, the 2 becomes -2 and the 3 becomes -3. Note that $-2 > -3$ (because -2 is closer to zero than -3 is), which shows that the $<$ sign turns into a $>$ in this case. (If you multiply or divide by a positive number, the direction doesn't change. If you add or subtract a negative number, the direction doesn't change.)

Example 1. $-8x > -16$. Divide by -8 and reverse the inequality: $x < 2$.
Note: We could get the same answer using more steps without dividing by a negative number. We could add $8x$ to get $0 > -16 + 8x$, add 16 to get $16 > 8x$, and then divide by 8 to get $2 > x$. Observe that $2 > x$ is equivalent to $x < 2$. (The vertex points to x in each case.)

Example 2. $x < 12 + 4x$. Subtract $4x$ from both sides: $-3x < 12$. Divide by -3 and reverse the inequality: $x > -4$.
Check: Again, we will check the answer using more steps. Subtract x from both sides: $0 < 12 + 3x$. Subtract 12 from both sides: $-12 < 3x$. Divide by 3: $-4 < x$. Observe that $-4 < x$ is equivalent to $x > -4$. (The vertex points to the -4 in each case.)

8.1 Problems

Directions: Isolate the unknown, dividing by a negative number in the last step.

(1) $-x < -11$

(2) $5 > -x$

---

(3) $-3x > 21$

(4) $-6x < -12$

(5) $5x \leq 9x + 24$

(6) $3x - 56 < -4x$

(7) $-4x - 20 > -9x$

(8) $35 + 8x \geq x$

(9) $-x < -1 + x$

(10) $8x - 100 < -2x$

(11) $-3(x - 5) > 6$

(12) $-6(x - 5) < -2(9 - x)$

(13) $9 + 5x + 6 \leq 2x$

(14) $-3x - 8x > -9x - 14$

8.2 Taking Square Roots of Both Sides of an Inequality

Recall from Sec. 2.9 that there are two solutions to consider when taking the square root of both sides of an equation. For example, the solutions to $x^2 = 4$ are $x = \pm 2$ because both roots satisfy the equation: $(-2)^2 = 4$ and $2^2 = 4$. There can similarly be two solutions when taking the square root of both sides of an inequality. For example, the solutions to $x^2 < 4$ include $x < 2$ as well as $x > -2$. Both answers can be combined together using absolute values: $|x| < 2$.

Although the equation $x^2 = -9$ only has the imaginary solutions $x = \pm 3i$ (Sec. 3.10), the inequality $x^2 > -9$ has all real numbers ($\mathbb{R}$) as its solution since any real number squared is nonnegative. In contrast, $x^2 < -9$ doesn't have any real solutions.

Example 1. $2x^2 > 32$. Divide by 2 on both sides: $x^2 > 16$. Square root both sides. Either $x > 4$ or $x < -4$. These two inequalities are equivalent to $|x| > 4$.

Example 2. $5x^2 + 4 < 2x^2 + 19$. Subtract 4 and $2x^2$ from both sides: $3x^2 < 15$. Divide by 3 on both sides: $x^2 < 5$. Square root both sides. Either $x < \sqrt{5}$ or $x > -\sqrt{5}$. These two inequalities are equivalent to $|x| < \sqrt{5}$.

Example 3. $x^2 > -2$. Any real number will satisfy this inequality: $x \in \mathbb{R}$.

8.2 Problems

Directions: Solve each inequality.

(1) $x^2 < 49$

(2) $6x^2 \geq 18$

(3) $20 < x^2 - 16$

(4) $5x^2 > -10$

(5) $7x^2 > 2x^2 + 20$

(6) $4x^2 - 21 < 43$

(7) $-8x^2 > -200$

(8) $24 - x^2 \leq x^2$

(9) $5 - 7x^2 < 9$

(10) $-4x^2 > -1$

(11) $9x^2 - 12 \geq 24 + 5x^2$

(12) $7 - 2x^2 < 2 - x^2$

(13) $-3x^2 - 83 > -6x^2 - 2$

(14) $4x^2 - 29 \leq -6x^2 + 71$

8.3 Quadratic Inequalities

The equation $(x - 2)(x - 3) = 0$ is true if $x - 2 = 0$ or if $x - 3 = 0$, which leads to two solutions: $x = 2$ or $x = 3$. With an inequality, it is a little different. For example, consider $(x - 2)(x - 3) > 0$. We are multiplying two numbers together: $(x - 2)$ and $(x - 3)$. When multiplying two numbers, there are two ways for the product to be positive. For example, consider that $(5)(4) = 20$ is positive and $(-5)(-4) = 20$ is also positive. This means that $(x - 2)(x - 3) > 0$ has two separate pairs of solutions:

- One possibility is that $x - 2 > 0$ and $x - 3 > 0$. These inequalities simplify to $x > 2$ and $x > 3$. The only way for both of these inequalities to be true is if $x > 3$.
- The other possibility is that $x - 2 < 0$ and $x - 3 < 0$. These inequalities simplify to $x < 2$ and $x < 3$. These can only both be true if $x < 2$.

The two possible solutions are $x < 2$ or $x > 3$. We can express this as $x \in (-\infty, 2) \cup (3, \infty)$, where $\cup$ is the symbol for the **union** of two sets, the symbol $\in$ means "is an **element** of," and $(-\infty, 2)$ and $(3, \infty)$ use interval notation (recall Sec. 1.2).

Now consider the inequality $(x - 2)(x - 3) < 0$. For the product to be negative, one number must be positive and one number must be negative. The possibilities are:

- $x - 2 > 0$ and $x - 3 < 0$, which simplify to $x > 2$ and $x < 3$. We can combine these together to get $2 < x < 3$.
- $x - 2 < 0$ and $x - 3 > 0$, which simplify to $x < 2$ and $x > 3$. No value of x will make <u>both</u> of these inequalities true.

The only viable solution is $2 < x < 3$, which can be expressed as $x \in (2, 3)$.

Example 1. $x^2 - 3x - 4 > 0$. First, factor the quadratic using the technique from Sec.'s 4.2-4.3: $(x + 1)(x - 4) > 0$. For the left-hand side to be positive, either both factors must be positive or both factors must be negative.

- One possibility is that $x + 1 > 0$ and $x - 4 > 0$, for which $x > -1$ and $x > 4$. The only way for both of these to be true is if $x > 4$.
- The other possibility is that $x + 1 < 0$ and $x - 4 < 0$, for which $x < -1$ and $x < 4$. The only way for both of these to be true is if $x < -1$.

Combine these answers: $x < -1$ or $x > 4$, which is equivalent to $x \in (-\infty, -1) \cup (4, \infty)$.

Example 2. $6x^2 + 5x - 6 < 0$. First, factor the quadratic using the technique from Sec. 4.3: $(2x + 3)(3x - 2) < 0$. For the left-hand side to be negative, one factor must be positive and one factor must be negative.

- One possibility is that $2x + 3 > 0$ and $3x - 2 < 0$, for which $x > -\frac{3}{2}$ and $x < \frac{2}{3}$. Combine these together to write $-\frac{3}{2} < x < \frac{2}{3}$.

- The other possibility is that $2x + 3 < 0$ and $3x - 2 > 0$, for which $x < -\frac{3}{2}$ and $x > \frac{2}{3}$. No value of x can make both of these inequalities true.

Final answer: $-\frac{3}{2} < x < \frac{2}{3}$, which is equivalent to $x \in \left(-\frac{3}{2}, \frac{2}{3}\right)$.

8.3 Problems

Directions: Find the complete solution to each inequality.

(1) $x^2 - 7x - 18 < 0$

(2) $x^2 - 13x + 40 > 0$

(3) $6x^2 + 5x - 4 > 0$

(4) $15x^2 + 34x + 15 < 0$

(5) $16x^2 - 46x + 15 > 0$

(6) $20x^2 - 9x - 20 < 0$

(7) $8x^2 + 49x + 45 < 0$

(8) $24x^2 - 2x - 15 > 0$

(9) $9x^2 - 42x + 49 > 0$

(10) $30x^2 + 42x - 36 < 0$

8.4 Taking Reciprocals of Both Sides of an Inequality

When taking the **reciprocal** of both sides, there are four cases to consider:

- $\frac{1}{x} > \frac{1}{5}$. Both sides must be positive. Reverse the direction of the inequality to get $x < 5$. Since x must be positive, the complete solution is $0 < x < 5$. (Why does the inequality reverse? You can see this the long way. Multiply by 5 on both sides to get $\frac{5}{x} > 1$. Then multiply by x on both sides: $5 > x$. Observe that $5 > x$ is equivalent to $x < 5$.)

- $\frac{1}{x} < \frac{1}{5}$. This time, the left-hand side is not necessarily positive. For positive x, reverse the direction of the inequality to get $x > 5$. Since all negative values of x satisfy the given inequality, the complete solution is $x < 0$ or $x > 5$.

- $\frac{1}{x} < -\frac{1}{5}$. Both sides must be negative. Reverse the direction of the inequality to get $x > -5$. Since x must be negative, the complete solution is $-5 < x < 0$.

- $\frac{1}{x} > -\frac{1}{5}$. This time, the left-hand side is not necessarily negative. For negative x, reverse the direction of the inequality to get $x < -5$. Since all positive values of x satisfy the given inequality, the complete solution is $x < -5$ or $x > 0$.

Example 1. $\frac{1}{x} > \frac{1}{9}$. Here, x must be less than 9 and positive: $0 < x < 9$. The answer may also be expressed as $x \in (0, 9)$. Recall Sec. 8.3.

Example 2. $\frac{1}{x} < \frac{1}{9}$. Here, x can be greater than 9 or negative: $x < 0$ or $x > 9$. The answer may also be expressed as $x \in (-\infty, 0) \cup (9, \infty)$.

Example 3. $\frac{1}{x} < -\frac{1}{9}$. Here, x must be greater than -9 and negative: $-9 < x < 0$. The answer may also be expressed as $x \in (-9, 0)$.

Example 4. $\frac{1}{x} > -\frac{1}{9}$. Here, x can be less than -9 or positive: $x < -9$ or $x > 0$. The answer may also be expressed as $x \in (-\infty, -9) \cup (0, \infty)$.

8.4 Problems

Directions: Find the complete solution to each inequality.

(1) $\frac{1}{x} > \frac{1}{6}$

(2) $\frac{1}{x} < \frac{1}{2}$

(3) $\frac{1}{x} < -\frac{1}{4}$

(4) $\frac{1}{x} > -\frac{1}{3}$

(5) $\frac{1}{8} > \frac{1}{x}$

(6) $\frac{1}{10} < \frac{1}{x}$

(7) $-\frac{1}{12} < \frac{1}{x}$

(8) $-\frac{1}{15} > \frac{1}{x}$

(9) $\frac{1}{x} > 3$

(10) $-\frac{1}{x} < \frac{1}{5}$

8.5 Cross Multiplying with Inequalities

<u>Cross multiplying</u> is equivalent to multiplying both sides of the equation by both of the denominators. For example, multiply by $6x$ on both sides of $\frac{4}{x} = \frac{8}{6}$ to get $24 = 8x$. When there is an inequality, if exactly one denominator is negative, the direction of the inequality reverses when you cross multiply, but if both denominators have the same sign, the direction remains unchanged. If there is a variable in a denominator, you must consider the possible signs of the variable, as shown in the examples.

Example 1. $\frac{6}{x} > \frac{9}{12}$. Here, x must be positive. When we cross multiply, the inequality remains unchanged: $6(12) > 9x$. Simplify: $72 > 9x$. Divide by 9 on both sides: $8 > x$. Since x must be positive, the answer is $0 < x < 8$, which is equivalent to $x \in (0, 8)$.

Example 2. $\frac{10}{x} > -\frac{15}{9}$. Here, any positive value for x satisfies the inequality. If x is negative, when we cross multiply, the direction of the inequality reverses because one denominator (x) is negative and the other denominator (9) is positive: $10(9) < -15x$. Simplify: $90 < -15x$. Divide by -15. The direction of the inequality reverses again when we divide by a negative number (Sec. 8.1): $-6 > x$. The complete solution is $x < -6$ or $x > 0$, which is equivalent to $x \in (-\infty, -6) \cup (0, \infty)$.

8.5 Problems

Directions: Find the complete solution to each inequality.

(1) $\frac{18}{x} > \frac{24}{20}$

(2) $\frac{16}{x} < \frac{10}{35}$

(3) $\frac{6}{x} > -\frac{3}{7}$

(4) $\frac{12}{x} < -\frac{72}{54}$

(5) $\frac{3}{8} > \frac{24}{x}$

(6) $\frac{27}{15} < \frac{36}{x}$

(7) $-\frac{39}{42} < \frac{26}{x}$

(8) $-\frac{11}{9} > \frac{121}{x}$

(9) $\frac{x}{9} > \frac{4}{x}$

(10) $-\frac{27}{x} < \frac{x}{3}$

ANSWER KEY

1 Functions

1.1 Problems

(1) Yes. For every value x for which f is real, there is exactly one value of f.

(2) No. For nonzero values of x, there are two values of f for each value of x. For example, when $x = 2$, f equals 2 and f also equals -2 because $f = \pm 2$.

(3) No. For $x = 6$, two different values of f are given (7 and 9).

(4) Yes. For every value of x in the table, there is exactly one value of f. (It's okay that $f = 5$ appears twice. It's whether or not f has multiple values for the same x that matters.)

(5) Yes. For every value of x in the graph, there is exactly one value of f.

(6) No. For each nonnegative value of x, there are two different values for f. This graph fails the **vertical line test**.

1.2 Problems

(1) Domain: $\mathbb{R}$. Range: $[0, \infty)$, which is equivalent to $f \geq 0$. Note: The square of a real number can't be negative.

(2) Domain: $[2, \infty)$, which is equivalent to $y \geq 2$. Range: $[0, \infty)$, which is equivalent to $g \geq 0$. Note: g equals its minimum value of zero when y is 2.

Note: Although $\sqrt{4}$ could technically be 2 or -2 because $2^2 = (2)(2) = 4$ and $(-2)^2 = (-2)(-2) = 4$, in algebra only the positive root is taken unless $\pm$ notation is used. That is, if we wanted $\sqrt{4}$ to mean both 2 and -2, we would write $\pm\sqrt{4}$. Here, $g(y)$ can't mean $\pm\sqrt{y-2}$ because then $g(y)$ wouldn't be a function (Sec. 1.1).

(3) Domain: $(-\infty, -3) \cup (-3, \infty)$, which is equivalent to $x \neq -3$. Range: $(-\infty, 0) \cup (0, \infty)$, which is equivalent to $y \neq 0$. **Note: $\cup$ is the symbol for the union of two sets.** It is equivalent to the word "or." The domain could also be expressed as $(-\infty, -3)$ or $(-3, \infty)$, meaning that x may be any value in the interval $(-\infty, -3)$ or in the interval $(-3, \infty)$. Both intervals exclude the value -3.

(4) Domain: $\mathbb{R}$. Range: $\mathbb{R}$.

(5) Domain: $\mathbb{R}$. Range: $[6, \infty)$, which is equivalent to $z \geq 6$. Note: Since $y^2 \geq 0$ and $y^4 \geq 0$, it follows that $y^4 + 3y^2 + 6 \geq 6$.

(6) Domain: $(-\infty, 0) \cup (0, \infty)$, which is equivalent to $u \neq 0$. Range: $(0, \infty)$, which is equivalent to $v > 0$. Notes: The square of a real number can't be negative. For finite values of u, v is always positive. Although v approaches zero as u approaches infinity, v never quite reaches exactly zero. See the note to the solution to Problem 3.

(7) Domain: $(-\infty, 4]$, which is equivalent to $t \leq 4$. Range: $[0, \infty)$, which is equivalent to $x \geq 0$. Note: x equals its minimum value of zero when t is 4.

(8) Domain: $\mathbb{R}$. Range: $[0, \infty)$, which is equivalent to $g \geq 0$. The square of a real number can't be negative. Note: g equals its minimum value of zero when x is -5.

(9) Domain: $(-\infty, -3) \cup (-3, 5) \cup (5, \infty)$, which is equivalent to $y \neq -3$ and $y \neq -5$. Range: $\left(-\infty, -\frac{7}{16}\right] \cup (0, \infty)$, which is equivalent to $f \leq -\frac{7}{16}$ or $f > 0$. Note: $(y + 3)(y - 5) = y^2 - 2y - 15$, which is a parabola. There are multiple ways to show that this parabola has a minimum value of -16 when y equals 1. For example, you can complete the square (as in Sec. 7.7) to write $y^2 - 2y - 15 = (y - 1)^2 - 16$. It follows that $f(y) = \frac{7}{y^2 - 2y - 15}$ can't lie between 0 and $-\frac{7}{16}$. See the note to the solution to Problem 3.

Tip: **All of these answers can be verified by making a graph**. We'll learn about graphs in Chapter 7. If you have a graphing calculator (or graph the function with an online program like Wolfram Alpha or Mathematica), you can also check the answers visually with the graph.

(10) Domain: $\mathbb{R}$. Range: $(-\infty, 9]$, which is equivalent to $z \leq 9$. Note: z equals its maximum value of 9 when w is zero.

(11) Domain: $(0, \infty)$, which is equivalent to $k > 0$. Unlike Example 1, this domain doesn't include 0 because the square root is in the denominator in this problem. Range: $(0, \infty)$, which is equivalent to $N > 0$.

(12) Domain: $[2, \infty)$, which is equivalent to $q \geq 2$. Range: $[0, \infty)$, which is equivalent to $p \geq 0$. Note: p equals its minimum value of zero when q is 2.

(13) Domain: $(-\infty, -1) \cup (-1, \infty)$, which is equivalent to $a \neq -1$. Range: $(-\infty, 1) \cup (1, \infty)$, which is equivalent to $b \neq 1$. See the note to the solution to Problem 3.

Notes: $b(a) = \frac{a}{a+1}$ is a hyperbola with two branches. When $a > -1$, $b < 1$ and when $a < -1$, $b > 1$. Here is one way to see that $b \neq 1$. Multiply both sides of $b = \frac{a}{a+1}$ by $a + 1$ to get $ba + b = a$, subtract ba from both sides to get $b = a - ba$, factor to get $b = a(1 - b)$, and divide by $1 - b$ to get $\frac{b}{1-b} = a$. Since the denominator of this equation is $1 - b$, this shows that $b \neq 1$ avoids a division by zero problem.

(14) Domain: $[0, \infty)$, which is equivalent to $r \geq 0$. Range: $[0, \infty)$, which is equivalent to $Q \geq 0$. Note: Q equals its minimum value of zero when r is zero.

Note: $r^{3/2} = |r^1|r^{1/2} = |r|\sqrt{r}$ because $r^1 = r$ and $r^{1/2} = \sqrt{r}$. See Chapter 2.

(15) Domain: $(-\infty, -3] \cup [3, \infty)$, which is equivalent to $t \leq -3$ or $t \geq 3$ (both inequalities are needed, though these may be condensed to the single inequality $t^2 \geq 9$, which itself is equivalent to $|t| \geq 3$). Range: $[0, \infty)$, which is equivalent to $T \geq 0$. Note: T equals its minimum value of zero when t is -3 or when t is 3. See the note to the solution to Problem 3.

(16) Domain: $(-\infty, -4) \cup (-4, 2) \cup (2, \infty)$, which is equivalent to $x \neq -4$ and $x \neq 2$.

Range: $\left(-\infty, -\frac{2}{9}\right] \cup (0, \infty)$, which is equivalent to $h \leq -\frac{2}{9}$ or $h > 0$.

Note: $x^2 + 2x - 8$ can be factored as $x^2 + 2x - 8 = (x + 4)(x - 2)$, which shows that h is a problem when x equals -4 or 2. The expression $x^2 + 2x - 8$ is a parabola. There are multiple ways to show that this parabola has a minimum value of -9 when x equals -1. For example, you can complete the square (as in Sec. 7.7) to write $x^2 + 2x - 8 = (x + 1)^2 - 9$. It follows that $h(x) = \frac{2}{x^2+2x-8}$ can't lie between 0 and $-\frac{2}{9}$. See the note to the solution to Problem 3.

1.3 Problems

(1) $f(3) = 3^2 - 3(3) + 8 = 9 - 9 + 8 = \boxed{8}$, $f(2) = 2^2 - 3(2) + 8 = 4 - 6 + 8 = \boxed{6}$
$f(1) = 1^2 - 3(1) + 8 = 1 - 3 + 8 = \boxed{6}$, $f(0) = 0^2 - 3(0) + 8 = 0 - 0 + 8 = \boxed{8}$
$f(-1) = (-1)^2 - 3(-1) + 8 = 1 + 3 + 8 = \boxed{12}$
(2) $g(2) = 7(2) - 5 = 14 - 5 = \boxed{9}$, $g(1) = 7(1) - 5 = 7 - 5 = \boxed{2}$
$g\left(\frac{1}{2}\right) = 7\left(\frac{1}{2}\right) - 5 = \frac{7}{2} - \frac{10}{2} = \boxed{-\frac{3}{2}} = \boxed{-1.5}$, $g\left(-\frac{1}{2}\right) = 7\left(-\frac{1}{2}\right) - 5 = -\frac{7}{2} - \frac{10}{2} = \boxed{-\frac{17}{2}} = \boxed{-8.5}$, $g(-1) = 7(-1) - 5 = -7 - 5 = \boxed{-12}$

(3) $y(5) = \sqrt{25 - 5^2} = \sqrt{25 - 25} = \boxed{0}$, $y(4) = \sqrt{25 - 4^2} = \sqrt{25 - 16} = \sqrt{9} = \boxed{3}$

$y(3) = \sqrt{25 - 3^2} = \sqrt{25 - 9} = \sqrt{16} = \boxed{4}$

$y(-1) = \sqrt{25 - (-1)^2} = \sqrt{25 - 1} = \sqrt{24} = \sqrt{(4)(6)} = \sqrt{4}\sqrt{6} = \boxed{2\sqrt{6}}$

$y(-3) = \sqrt{25 - (-3)^2} = \sqrt{25 - 9} = \sqrt{16} = \boxed{4}$

(4) $v(4) = \frac{4}{3} - \frac{4}{4} = \frac{4}{3} - 1 = \frac{4}{3} - \frac{3}{3} = \boxed{\frac{1}{3}}$, $v(3) = \frac{3}{3} - \frac{4}{3} = \boxed{-\frac{1}{3}}$

$v(2) = \frac{2}{3} - \frac{4}{2} = \frac{2}{3} - 2 = \frac{2}{3} - \frac{6}{3} = \boxed{-\frac{4}{3}}$, $v(1) = \frac{1}{3} - \frac{4}{1} = \frac{1}{3} - \frac{12}{3} = \boxed{-\frac{11}{3}}$

$v(-12) = -\frac{12}{3} - \frac{4}{-12} = -\frac{12}{3} + \frac{1}{3} = \boxed{-\frac{11}{3}}$

(5) $f(27) = 27^{2/3} = \left(27^{1/3}\right)^2 = 3^2 = \boxed{9}$, $f(8) = 8^{2/3} = \left(8^{1/3}\right)^2 = 2^2 = \boxed{4}$

$f(-8) = (-8)^{2/3} = \left[(-8)^{1/3}\right]^2 = (-2)^2 = (-2)(-2) = \boxed{4}$

$f(-27) = (-27)^{2/3} = \left[(-27)^{1/3}\right]^2 = (-3)^2 = (-3)(-3) = \boxed{9}$

Note: Recall the rule $x^{ab} = (x^a)^b = (x^b)^a$. Let $b = \frac{1}{c}$ to see that $x^{a/c} = (x^a)^{1/c} = \left(x^{1/c}\right)^a$. If needed, you can also check the answer using a calculator. To enter $27^{2/3}$ on most calculators, type 27^(2/3). To enter $(-27)^{2/3}$, type (−27)^(2/3).

1.4 Problems

(1) $f(3, 2) = 2^2 - 3^2 = 4 - 9 = \boxed{-5}$. Note: $x = 3$ and $y = 2$.

(2) $g(6, 54) = \sqrt{6(54)} = \sqrt{6(6)(9)} = \sqrt{36}\sqrt{9} = (6)(3) = \boxed{18}$.

(3) $Q(4, 3) = \frac{1}{4} + \frac{4}{3} = \frac{3}{12} + \frac{16}{12} = \boxed{\frac{19}{12}}$.

(4) $d(3, 4, 5) = 5^2 - (3)(4) = 25 - 12 = \boxed{13}$. Note: $a = 3$, $b = 4$, and $c = 5$.

1.5 Problems

(1) $f(-3) = (-3)^2 - 9 = 9 - 9 = \boxed{0}$, $f(0) = 0^2 - 9 = \boxed{-9}$

$f(2) = 2^2 - 9 = 4 - 9 = \boxed{-5}$, $f(3) = 3^2 - 9 = 9 - 9 = \boxed{0}$, $f(5) = 3 - 5 = \boxed{-2}$

(2) $g\left(-\frac{1}{4}\right) = 1 - \left(-\frac{1}{4}\right) = 1 + \frac{1}{4} = \frac{4}{4} + \frac{1}{4} = \boxed{\frac{5}{4}}$, $g(0) = 2(0) + 1 = \boxed{1}$

$g(1) = 2(1) + 1 = 2 + 1 = \boxed{3}$, $g\left(\frac{3}{2}\right) = 2\left(\frac{3}{2}\right) + 1 = 3 + 1 = \boxed{4}$, and $g(2) = 2^2 = \boxed{4}$

(3) $f(-3) = |3 - 2(-3)| = |3 + 6| = \boxed{9}$, $f(0) = |3 - 2(0)| = \boxed{3}$

$f(1) = |3 - 2(1)| = |3 - 2| = \boxed{1}$, $f(2) = |3 - 2(2)| = |3 - 4| = |-1| = \boxed{1}$

$f(3) = |3 - 2(3)| = |3 - 6| = |-3| = \boxed{3}$

(4) $g(-3) = |2 - (-3)^2| = |2 - 9| = |-7| = \boxed{7}$, $g(-1) = |2 - (-1)^2| = |2 - 1| = \boxed{1}$

$g(0) = |2 - 0^2| = \boxed{2}$, $g(1) = |2 - 1^2| = |2 - 1| = \boxed{1}$

$g(3) = |2 - 3^2| = |2 - 9| = |-7| = \boxed{7}$

(5) $h(0) = |0 - \sqrt{0}| = \boxed{0}$, $h\left(\frac{1}{4}\right) = \left|\frac{1}{4} - \sqrt{\frac{1}{4}}\right| = \left|\frac{1}{4} - \frac{1}{2}\right| = \left|\frac{1}{4} - \frac{2}{4}\right| = \left|-\frac{1}{4}\right| = \boxed{\frac{1}{4}}$

$h(1) = |1 - \sqrt{1}| = |1 - 1| = \boxed{0}$, $h\left(\frac{9}{4}\right) = \left|\frac{9}{4} - \sqrt{\frac{9}{4}}\right| = \left|\frac{9}{4} - \frac{3}{2}\right| = \left|\frac{9}{4} - \frac{6}{4}\right| = \boxed{\frac{3}{4}}$

$h(4) = |4 - \sqrt{4}| = |4 - 2| = \boxed{2}$

1.6 Problems

(1) $g(6) = 4(6) + 3 = 24 + 3 = 27$, $f\big(g(6)\big) = f(27) = 5(27) - 2 = 135 - 2 = \boxed{133}$

$f(6) = 5(6) - 2 = 30 - 2 = 28$, $g\big(f(6)\big) = g(28) = 4(28) + 3 = 112 + 3 = \boxed{115}$

$f\big(g(x)\big) = f(4x + 3) = 5(4x + 3) - 2 = 20x + 15 - 2 = \boxed{20x + 13}$

$g\big(f(x)\big) = g(5x - 2) = 4(5x - 2) + 3 = 20x - 8 + 3 = \boxed{20x - 5}$

(2) $g(1) = 1^2 + 3 = 1 + 3 = 4$, $f\big(g(1)\big) = f(4) = \frac{24}{4} = \boxed{6}$

$g(3) = 3^2 + 3 = 9 + 3 = 12$, $f\big(g(3)\big) = f(12) = \frac{24}{12} = \boxed{2}$, $f\big(g(t)\big) = f(t^2 + 3) = \boxed{\frac{24}{t^2+3}}$

(3) $f(1) = \sqrt{7(1) + 2} = \sqrt{9} = 3$, $f\big(f(1)\big) = f(3) = \sqrt{7(3) + 2} = \boxed{\sqrt{23}}$

$f(2) = \sqrt{7(2) + 2} = \sqrt{16} = 4$, $f\big(f(2)\big) = f(4) = \sqrt{7(4) + 2} = \boxed{\sqrt{30}}$

$f(14) = \sqrt{7(14) + 2} = \sqrt{100} = 10$, $f\big(f(14)\big) = f(10) = \sqrt{7(10) + 2} = \sqrt{72} = \boxed{6\sqrt{2}}$

$f\big(f(x)\big) = f(\sqrt{7x + 2}) = \boxed{\sqrt{7\sqrt{7x + 2} + 2}}$

Notes: We factored out a perfect square in order to express $\sqrt{72}$ in standard form: $\sqrt{72} = \sqrt{(36)(2)} = \sqrt{36}\sqrt{2} = 6\sqrt{2}$. If you plug 1, 2, or 14 into $f\big(f(x)\big) = \sqrt{7\sqrt{7x + 2} + 2}$, you will see that this agrees with the answers for $f\big(f(1)\big)$, $f\big(f(2)\big)$, and $f\big(f(14)\big)$.

(4) $f(3) = 3^2 - 4 = 9 - 4 = 5, f(f(3)) = f(5) = 5^2 - 4 = 25 - 4 = \boxed{21}$

$g(3) = 3 + 8 = 11, f(g(3)) = f(11) = 11^2 - 4 = 121 - 4 = \boxed{117}$

$f(3) = 3^2 - 4 = 9 - 4 = 5, g(f(3)) = g(5) = 5 + 8 = \boxed{13}$

$g(3) = 3 + 8 = 11, g(g(3)) = g(11) = 11 + 8 = \boxed{19}$

(5) $q(4) = \frac{12}{4} = 3, p(q(4)) = p(3) = -8 - 3^2 = -8 - 9 = \boxed{-17}$

$p(4) = -8 - 4^2 = -8 - 16 = -24, q(p(4)) = q(-24) = \frac{12}{-24} = \boxed{-\frac{1}{2}} = \boxed{-0.5}$

(6) $g(h(x)) = 4(4x + 3) - 5 = 16x + 12 - 5 = 16x + 7$

$f(g(h(x))) = f(16x + 7) = 3(16x + 7) + 2 = 48x + 21 + 2 = \boxed{48x + 23}$

1.7 Problems

(1) $5f^{-1} - 4 = x$

$5f^{-1} = x + 4$

$f^{-1}(x) = \boxed{\frac{x+4}{5}}$

(2) $8 - 7g^{-1} = x$

$-7g^{-1} = x - 8$

$g^{-1}(x) = \frac{x-8}{-7} = \boxed{\frac{8-x}{7}}$

(3) $\boxed{\text{no inverse}}$

$h(-2) = (-2)^4 = 16$ equals

$h(2) = 2^4 = 16$

Note: Multiply the numerator and denominator of $\frac{x-8}{-7}$ by -1 to see that $\frac{x-8}{-7} = \frac{8-x}{7}$.

(4) $(p^{-1})^5 = y$

$p^{-1}(y) = \boxed{y^{1/5}} = \boxed{\sqrt[5]{y}}$

(5) $\frac{6}{f^{-1}} = x$

$6 = xf^{-1}$

$f^{-1}(x) = \boxed{\frac{6}{x}}$

(6) $\frac{9}{2q^{-1}+5} = t$

$9 = 2tq^{-1} + 5t$

$9 - 5t = 2tq^{-1}$

$q^{-1}(t) = \boxed{\frac{9-5t}{2t}} = \boxed{\frac{9}{2t} - \frac{5}{2}}$

Note: To solve for p^{-1} in Problem 4, take the fifth root of both sides. For example, the solution to $x^5 = 32$ is $x = 32^{1/5} = \sqrt[5]{32} = 2$. Check: $2^5 = 32$.

(7) $\frac{g^{-1}+2}{g^{-1}-3} = x$

$g^{-1} + 2 = xg^{-1} - 3x$

$3x + 2 = xg^{-1} - g^{-1}$

$3x + 2 = g^{-1}(x - 1)$

$g^{-1}(x) = \boxed{\frac{3x+2}{x-1}}$

(8) $\frac{4h^{-1}-7}{5h^{-1}+8} = k$

$4h^{-1} - 7 = 5kh^{-1} + 8k$

$4h^{-1} - 5kh^{-1} = 8k + 7$

$h^{-1}(4 - 5k) = 8k + 7$

$h^{-1}(k) = \boxed{\frac{8k+7}{4-5k}} = \boxed{\frac{8k+7}{-5k+4}}$

(9) $\sqrt{3f^{-1}+8}=y$

$3f^{-1}+8=y^2$

$3f^{-1}=y^2-8$

$f^{-1}=\frac{y^2-8}{3}$

no inverse

(10) $f(x)=x^2-4x-12$

$f(x)=(x-6)(x+2)$

$f(6)=0$ and $f(-2)=0$

no inverse

Note: In Problem 9, $\frac{y^2-8}{3}$ is NOT the inverse of $f(y)$. Why? In order for $g(y)=\frac{y^2-8}{3}$

and $f(y)=\sqrt{3y+8}$ to be inverse functions, $g(y)=\frac{y^2-8}{3}$ and $f(y)=\sqrt{3y+8}$ **both**

need to be inverses of each other. However, $g(y)=\frac{y^2-8}{3}$ does NOT have an inverse.

Note that $g(-1)=-\frac{7}{3}$ and $g(1)=-\frac{7}{3}$, which shows that $g(y)$ has no inverse. If we

restrict the domain of $g(y)$ to $y\geq 0$ or $y\leq 0$, then for this restricted domain, $\frac{y^2-8}{3}$

would be the inverse of $f(y)$.

1.8 Problems

(1) Odd: $f(-x)=(-x)^5=-x^5=-f(x)$

(2) Even: $g(-x)=(-x)^4=x^4=g(x)$

(3) Even: $h(-t)=(-t)^8-5(-t)^6-3=t^8-5t^6-3=h(t)$

(4) Odd: $p(-y)=-4(-y)^9+2(-y)^5-(-y)=4y^9-2y^5+y=-p(y)$

(5) Neither: $q(-r)=3(-r)+8=-3r+8$

Note: $q(r)$ is a polynomial that has an odd term and a constant term.

(6) Neither: $f(-x)=\sqrt{-x}$ (which is imaginary when $x>0$)

(7) Even: $g(-y)=\sqrt{(-y)^2-9}=\sqrt{y^2-9}=g(y)$

(8) Odd: $v(-t)=\frac{5}{(-t)^3-6(-t)}=\frac{5}{-t^3+6t}=-v(t)$

(9) Even: $f(x)=(x+7)(x-7)=x^2-49, f(-x)=(-x)^2-49=x^2-49=f(x)$

(10) Neither: $q(z)=(z-3)(z+6)=z^2+3z-18$

Note: $q(z)$ is a polynomial with both even and odd terms.

(11) Odd: $f(-y)=\sqrt[3]{-y}=-\sqrt[3]{y}=-f(y)$

(12) Neither: $y(-x)=4^{-x}=\frac{4}{x}$

2 Radicals

2.1 Problems

(1) $\left(\sqrt{9x}\right)^2 = \boxed{9x}$ 	(2) $\sqrt{3}\sqrt{3} = \boxed{3}$

(3) $\sqrt{5x-4}\sqrt{5x-4} = \boxed{5x-4}$ 	(4) $\left(\sqrt{x+6}\right)^2 = \boxed{x+6}$

(5) $\left(\sqrt{5x}\right)^4 = (5x)^2 = 5^2x^2 = \boxed{25x^2}$ 	(6) $\left(\sqrt{3x}\right)^3 = \left(\sqrt{3x}\right)^2\sqrt{3x} = \boxed{3x\sqrt{3x}}$

(7) $\left(\sqrt{x-9}\right)^8 = \boxed{(x-9)^4} = \boxed{x^4 - 36x^3 + 486x^2 - 2916x + 6561}$

(8) $\left(\sqrt{2x+7}\right)^5 = \left(\sqrt{2x+7}\right)^4\sqrt{2x+7} = \boxed{(2x+7)^2\sqrt{2x+7}} = \boxed{(4x^2 + 28x + 49)\sqrt{2x+7}}$

(9) $\left(\sqrt{5}\right)^4\left(\sqrt{10x}\right)^3 = 5^2\left(\sqrt{10x}\right)^2\sqrt{10x} = 25(10x)^1\sqrt{10x} = 25(10x)\sqrt{10x} = \boxed{250x\sqrt{10x}}$

(10) $\left(\sqrt{x+6}\right)^4\left(\sqrt{x-6}\right)^5 = (x+6)^2\left(\sqrt{x-6}\right)^4\sqrt{x-6} = (x+6)^2(x-6)^2\sqrt{x-6} =$

$[(x+6)(x-6)]^2\sqrt{x-6} = \boxed{(x^2-36)^2\sqrt{x-6}} = \boxed{(x^4 - 72x^2 + 1296)\sqrt{x-6}}$

Note: We applied the rule $v^2w^2 = (vw)^2$ with $v = x+6$ and $w = x-6$.

(11) $\sqrt{x}\sqrt{x}\sqrt{x}\sqrt{x} = xx = \boxed{x^2}$

(12) $\sqrt{x+2}\sqrt{x+2}\sqrt{x-2}\sqrt{x-2} = (x+2)(x-2) = \boxed{x^2 - 4}$

(13) $\sqrt{x-4}\sqrt{2x+1}\sqrt{x-4}\sqrt{2x+1} = (x-4)(2x+1) = \boxed{2x^2 - 7x - 4}$

(14) $\left(\sqrt{3x}\right)^4\left(\sqrt{9x}\right)^2 = (3x)^2(9x) = (9x^2)(9x) = \boxed{81x^3}$

(15) $\left(\sqrt{8x}\right)^3\sqrt{x} = \left(\sqrt{8x}\right)^2\sqrt{8x}\sqrt{x} = (8x)\sqrt{8}\sqrt{x}\sqrt{x} = 8x\sqrt{4}\sqrt{2}(x) = 8x^2(2)\sqrt{2} = \boxed{16x^2\sqrt{2}}$

Note: We applied the rule $\sqrt{vw} = \sqrt{v}\sqrt{w}$ to write $\sqrt{8x} = \sqrt{8}\sqrt{x}$. Also, $\sqrt{8} = \sqrt{4}\sqrt{2} = 2\sqrt{2}$.

(16) $\left(\sqrt{7x}\right)^5\left(\sqrt{2x}\right)^3 = \left(\sqrt{7x}\right)^4\sqrt{7x}\left(\sqrt{2x}\right)^2\sqrt{2x} = (7x)^2\sqrt{7x}(2x)\sqrt{2x}$

$= (49x^2)(2x)\sqrt{7}\sqrt{x}\sqrt{2}\sqrt{x} = 98x^3\sqrt{14}(x) = \boxed{98x^4\sqrt{14}}$

(17) $\left(\sqrt{2x}\right)^6\left(\sqrt{6x}\right)^4\left(\sqrt{3x}\right)^2 = (2x)^3(6x)^2(3x) = (8x^3)(36x^2)(3x) = \boxed{864x^6}$

(18) $\left(\sqrt{3x}\right)^5\left(\sqrt{6x}\right)^3\left(\sqrt{2x}\right) = \left(\sqrt{3x}\right)^4\sqrt{3x}\left(\sqrt{6x}\right)^2\sqrt{6x}\sqrt{2x} = (3x)^2\sqrt{3x}(6x)^1\sqrt{6x}\sqrt{2x}$

$= (9x^2)\sqrt{3}\sqrt{x}(6x)\sqrt{6}\sqrt{x}\sqrt{2}\sqrt{x} = (54x^3)\sqrt{x}\sqrt{x}\sqrt{x}\sqrt{3}\sqrt{6}\sqrt{2} = 54x^3(x)\sqrt{x}\sqrt{(3)(2)}\sqrt{6}$

$= 54x^4\sqrt{x}\sqrt{6}\sqrt{6} = (6)(54)x^4\sqrt{x} = \boxed{324x^4\sqrt{x}}$

(19) $\left(\sqrt{5x}\right)^4\left(\sqrt{x}\right)^3\left(\sqrt{7x}\right)^2 = (5x)^2\left(\sqrt{x}\right)^2\sqrt{x}(7x)^1 = (25x^2)(x)\sqrt{x}(7x) = \boxed{175x^4\sqrt{x}}$

(20) $\left(\sqrt{6x}\right)^5\left(\sqrt{7x}\right)^4\left(\sqrt{2x}\right)^3 = \left(\sqrt{6x}\right)^4\sqrt{6x}(7x)^2\left(\sqrt{2x}\right)^2\sqrt{2x} = (6x)^2\sqrt{6x}(49x^2)(2x)\sqrt{2x}$

$= (36x^2)\sqrt{6x}(98x^3)\sqrt{2x} = 3528x^5\sqrt{6}\sqrt{x}\sqrt{2}\sqrt{x} = 3528x^5(x)\sqrt{3}\sqrt{2}\sqrt{2} = \boxed{7056x^6\sqrt{3}}$

2.2 Problems

(1) $\sqrt{50} = \sqrt{(50)(2)} = \sqrt{25}\sqrt{2} = \boxed{5\sqrt{2}}$

(2) $\sqrt{45x^4} = \sqrt{(9)(5)}\sqrt{x^4} = x^2\sqrt{9}\sqrt{5} = \boxed{3x^2\sqrt{5}}$

(3) $\sqrt{12x^7} = \sqrt{(4)(3)x^6 x} = \sqrt{4}\sqrt{x^6}\sqrt{3x} = \boxed{2x^3\sqrt{3x}}$

(4) $\sqrt{98x^8} = \sqrt{(49)(2)}\sqrt{x^8} = x^4\sqrt{49}\sqrt{2} = \boxed{7x^4\sqrt{2}}$

(5) $\sqrt{63x^2} = \sqrt{(9)(7)}\sqrt{x^2} = x\sqrt{9}\sqrt{7} = \boxed{3x\sqrt{7}}$

(6) $\sqrt{192x^{11}} = \sqrt{(64)(3)x^{10}x} = \sqrt{64}\sqrt{x^{10}}\sqrt{3x} = \boxed{8x^5\sqrt{3x}}$

(7) $\sqrt{1100x^5} = \sqrt{(100)(11)x^4 x} = \sqrt{100}\sqrt{x^4}\sqrt{11x} = \boxed{10x^2\sqrt{11x}}$

(8) $\sqrt{288x^{16}} = \sqrt{(144)(2)}\sqrt{x^{16}} = x^8\sqrt{144}\sqrt{2} = \boxed{12x^8\sqrt{2}}$

(9) $\sqrt{567x^6} = \sqrt{(81)(7)}\sqrt{x^6} = x^3\sqrt{81}\sqrt{7} = \boxed{9x^3\sqrt{7}}$

(10) $\sqrt{216x^9} = \sqrt{(36)(6)x^8 x} = \sqrt{36}\sqrt{x^8}\sqrt{6x} = \boxed{6x^4\sqrt{6x}}$

2.3 Problems

(1) $\sqrt[4]{162x^8} = \sqrt[4]{81}\sqrt[4]{2}\sqrt[4]{x^8} = \boxed{3x^2\sqrt[4]{2}}$ Note that $(x^2)^4 = x^8$.

(2) $\sqrt[3]{24x^{12}} = \sqrt[3]{8}\sqrt[3]{3}\sqrt[3]{x^{12}} = \boxed{2x^4\sqrt[3]{3}}$ Note that $(x^4)^3 = x^{12}$.

(3) $\sqrt[5]{160x^{14}} = \sqrt[5]{32}\sqrt[5]{5}\sqrt[5]{x^{10}}\sqrt[5]{x^4} = \boxed{2x^2\sqrt[5]{5x^4}}$ Note that $(x^2)^5 = x^{10}$.

(4) $\sqrt[4]{80x^7} = \sqrt[4]{16}\sqrt[4]{5}\sqrt[4]{x^4}\sqrt[4]{x^3} = \boxed{2x\sqrt[4]{5x^3}}$ Note: The x out front is **not** raised to a power.

(5) $\sqrt[3]{250x^{10}} = \sqrt[3]{125}\sqrt[3]{2}\sqrt[3]{x^9}\sqrt[3]{x} = \boxed{5x^3\sqrt[3]{2x}}$ Note that $(x^3)^3 = x^9$.

(6) $\sqrt[6]{192x^{27}} = \sqrt[6]{64}\sqrt[6]{3}\sqrt[6]{x^{24}}\sqrt[6]{x^3} = \boxed{2x^4\sqrt[6]{3x^3}}$ Note that $(x^4)^6 = x^{24}$.

(7) $\sqrt[4]{512x^9} = \sqrt[4]{256}\sqrt[4]{2}\sqrt[4]{x^8}\sqrt[4]{x} = \boxed{4x^2\sqrt[4]{2x}}$ Note that $(x^2)^4 = x^8$.

(8) $\sqrt[3]{-81x^5} = \sqrt[3]{(-27)}\sqrt[3]{3}\sqrt[3]{x^3}\sqrt[3]{x^2} = \boxed{-3x\sqrt[3]{3x^2}}$ Note: The x out front has **no** exponent.

(9) $\sqrt[5]{-243x^{11}} = \sqrt[5]{(-243)}\sqrt[5]{x^{10}}\sqrt[5]{x} = \boxed{-3x^2\sqrt[5]{x}}$ Note that $(x^2)^5 = x^{10}$.

(10) $\sqrt[4]{70{,}000x^{18}} = \sqrt[4]{10{,}000}\sqrt[4]{7}\sqrt[4]{x^{16}}\sqrt[4]{x^2} = \boxed{10x^4\sqrt[4]{7x^2}}$ Note that $(x^4)^4 = x^{16}$.

2.4 Problems

(1) $\dfrac{1}{\sqrt{2}} = \dfrac{1}{\sqrt{2}}\dfrac{\sqrt{2}}{\sqrt{2}} = \boxed{\dfrac{\sqrt{2}}{2}}$

(2) $\dfrac{6}{\sqrt{3}} = \dfrac{6}{\sqrt{3}}\dfrac{\sqrt{3}}{\sqrt{3}} = \dfrac{6\sqrt{3}}{3} = \boxed{2\sqrt{3}}$

(3) $\dfrac{x}{2\sqrt{7}} = \dfrac{x}{2\sqrt{7}}\dfrac{\sqrt{7}}{\sqrt{7}} = \dfrac{x\sqrt{7}}{2(7)} = \boxed{\dfrac{x\sqrt{7}}{14}}$

(4) $\dfrac{24x}{\sqrt{6}} = \dfrac{24x}{\sqrt{6}}\dfrac{\sqrt{6}}{\sqrt{6}} = \dfrac{24x\sqrt{6}}{6} = \boxed{4x\sqrt{6}}$

(5) $\dfrac{1}{\sqrt{x}} = \dfrac{1}{\sqrt{x}}\dfrac{\sqrt{x}}{\sqrt{x}} = \boxed{\dfrac{\sqrt{x}}{x}}$

(6) $\dfrac{5}{\sqrt{10x}} = \dfrac{5}{\sqrt{10x}}\dfrac{\sqrt{10x}}{\sqrt{10x}} = \dfrac{5\sqrt{10x}}{10x} = \boxed{\dfrac{\sqrt{10x}}{2x}}$

(7) $\dfrac{9x}{\sqrt{3x}} = \dfrac{9x}{\sqrt{3x}}\dfrac{\sqrt{3x}}{\sqrt{3x}} = \dfrac{9x\sqrt{3x}}{3x} = \boxed{3\sqrt{3x}}$

(8) $\dfrac{1}{x\sqrt{x}} = \dfrac{1}{x\sqrt{x}}\dfrac{\sqrt{x}}{\sqrt{x}} = \dfrac{\sqrt{x}}{x(x)} = \boxed{\dfrac{\sqrt{x}}{x^2}}$

(9) $\dfrac{\sqrt{3x}}{\sqrt{6}} = \dfrac{\sqrt{3x}}{\sqrt{6}}\dfrac{\sqrt{6}}{\sqrt{6}} = \dfrac{\sqrt{18x}}{6} = \dfrac{\sqrt{(9)(2x)}}{6} = \dfrac{\sqrt{9}\sqrt{2x}}{6} = \dfrac{3\sqrt{2x}}{6} = \boxed{\dfrac{\sqrt{2x}}{2}}$

(10) $\dfrac{\sqrt{5}}{\sqrt{15x}} = \dfrac{\sqrt{5}}{\sqrt{15x}}\dfrac{\sqrt{15x}}{\sqrt{15x}} = \dfrac{\sqrt{75x}}{15x} = \dfrac{\sqrt{(25)(3x)}}{15x} = \dfrac{\sqrt{25}\sqrt{3x}}{15x} = \dfrac{5\sqrt{3x}}{15x} = \boxed{\dfrac{\sqrt{3x}}{3x}}$

(11) $\dfrac{x^2}{2\sqrt{x}} = \dfrac{x^2}{2\sqrt{x}}\dfrac{\sqrt{x}}{\sqrt{x}} = \dfrac{x^2\sqrt{x}}{2x} = \boxed{\dfrac{x\sqrt{x}}{2}}$

(12) $\dfrac{\sqrt{35}}{\sqrt{21}} = \dfrac{\sqrt{35}}{\sqrt{21}}\dfrac{\sqrt{21}}{\sqrt{21}} = \dfrac{\sqrt{(35)(21)}}{21} = \dfrac{\sqrt{(7)(5)(7)(3)}}{21} = \dfrac{\sqrt{49}\sqrt{15}}{21} = \dfrac{7\sqrt{15}}{21} = \boxed{\dfrac{\sqrt{15}}{3}}$

2.5 Problems

(1) $\dfrac{2+\sqrt{3}}{2-\sqrt{3}} = \dfrac{2+\sqrt{3}}{2-\sqrt{3}}\left(\dfrac{2+\sqrt{3}}{2+\sqrt{3}}\right) = \dfrac{2(2)+2\sqrt{3}+2\sqrt{3}+\sqrt{3}\sqrt{3}}{2(2)+2\sqrt{3}-2\sqrt{3}-\sqrt{3}\sqrt{3}} = \dfrac{4+4\sqrt{3}+3}{4-3} = \dfrac{7+4\sqrt{3}}{1} = \boxed{7+4\sqrt{3}}$

(2) $\dfrac{\sqrt{10}}{2+\sqrt{5}} = \dfrac{\sqrt{10}}{2+\sqrt{5}}\left(\dfrac{2-\sqrt{5}}{2-\sqrt{5}}\right) = \dfrac{2\sqrt{10}-\sqrt{10}\sqrt{5}}{2(2)-2\sqrt{5}+2\sqrt{5}-\sqrt{5}\sqrt{5}} = \dfrac{2\sqrt{10}-\sqrt{50}}{4-5} = \dfrac{2\sqrt{10}-\sqrt{25}\sqrt{2}}{-1} = \boxed{-2\sqrt{10}+5\sqrt{2}}$

(3) $\dfrac{3-\sqrt{7}}{4-\sqrt{7}} = \dfrac{3-\sqrt{7}}{4-\sqrt{7}}\left(\dfrac{4+\sqrt{7}}{4+\sqrt{7}}\right) = \dfrac{3(4)+3\sqrt{7}-4\sqrt{7}-\sqrt{7}\sqrt{7}}{4(4)+4\sqrt{7}-4\sqrt{7}-\sqrt{7}\sqrt{7}} = \dfrac{12-\sqrt{7}-7}{16-7} = \boxed{\dfrac{5-\sqrt{7}}{9}}$

(4) $\dfrac{3+\sqrt{6}}{9+\sqrt{2}} = \dfrac{3+\sqrt{6}}{9+\sqrt{2}}\left(\dfrac{9-\sqrt{2}}{9-\sqrt{2}}\right) = \dfrac{3(9)-3\sqrt{2}+9\sqrt{6}-\sqrt{6}\sqrt{2}}{9(9)-9\sqrt{2}+9\sqrt{2}-\sqrt{2}\sqrt{2}} = \dfrac{27-3\sqrt{2}+9\sqrt{6}-\sqrt{3}\sqrt{2}\sqrt{2}}{81-2} = \boxed{\dfrac{27-3\sqrt{2}+9\sqrt{6}-2\sqrt{3}}{79}}$

(5) $\dfrac{\sqrt{3}+\sqrt{2}}{\sqrt{3}-\sqrt{2}} = \dfrac{\sqrt{3}+\sqrt{2}}{\sqrt{3}-\sqrt{2}}\left(\dfrac{\sqrt{3}+\sqrt{2}}{\sqrt{3}+\sqrt{2}}\right) = \dfrac{\sqrt{3}\sqrt{3}+\sqrt{3}\sqrt{2}+\sqrt{2}\sqrt{3}+\sqrt{2}\sqrt{2}}{\sqrt{3}\sqrt{3}+\sqrt{3}\sqrt{2}-\sqrt{2}\sqrt{3}-\sqrt{2}\sqrt{2}} = \dfrac{3+2\sqrt{6}+2}{3-2} = \dfrac{5+2\sqrt{6}}{1} = \boxed{5+2\sqrt{6}}$

(6) $\frac{x-\sqrt{x}}{x+\sqrt{x}} = \frac{x-\sqrt{x}}{x+\sqrt{x}}\left(\frac{x-\sqrt{x}}{x-\sqrt{x}}\right) = \frac{x^2-x\sqrt{x}-x\sqrt{x}+\sqrt{x}\sqrt{x}}{x^2-x\sqrt{x}+x\sqrt{x}-\sqrt{x}\sqrt{x}} = \frac{x^2-2x\sqrt{x}+x}{x^2-x} = \frac{x(x-2\sqrt{x}+1)}{x(x-1)} = \boxed{\frac{x-2\sqrt{x}+1}{x-1}}$ if

$x \neq 0$ and if $x \neq 1$. The special case $x = 0$ is indeterminate. For the special case $x = 1$, the answer is zero because $\frac{x-\sqrt{x}}{x+\sqrt{x}}$ equals $\frac{1-\sqrt{1}}{1+\sqrt{1}} = \frac{1-1}{1+1} = \frac{0}{2} = 0$ for this value of x.

(7) $\frac{1+\sqrt{x}}{2-\sqrt{x}} = \frac{1+\sqrt{x}}{2-\sqrt{x}}\left(\frac{2+\sqrt{x}}{2+\sqrt{x}}\right) = \frac{1(2)+\sqrt{x}+2\sqrt{x}+\sqrt{x}\sqrt{x}}{2(2)+2\sqrt{x}-2\sqrt{x}-\sqrt{x}\sqrt{x}} = \frac{2+3\sqrt{x}+x}{4-x} = \boxed{\frac{x+3\sqrt{x}+2}{-x+4}}$

(8) $\frac{x-\sqrt{6}}{x+\sqrt{3}} = \frac{x-\sqrt{6}}{x+\sqrt{3}}\left(\frac{x-\sqrt{3}}{x-\sqrt{3}}\right) = \frac{x^2-x\sqrt{3}-x\sqrt{6}+\sqrt{6}\sqrt{3}}{x^2-x\sqrt{3}+x\sqrt{3}-3} = \frac{x^2-x\sqrt{3}-x\sqrt{6}+\sqrt{2}\sqrt{3}\sqrt{3}}{x^2-3} = \boxed{\frac{x^2-x\sqrt{6}-x\sqrt{3}+3\sqrt{2}}{x^2-3}}$

(9) $\frac{\sqrt{x}}{\sqrt{x}-1} = \frac{\sqrt{x}}{\sqrt{x}-1}\left(\frac{\sqrt{x}+1}{\sqrt{x}+1}\right) = \frac{\sqrt{x}\sqrt{x}+\sqrt{x}}{\sqrt{x}\sqrt{x}+\sqrt{x}-\sqrt{x}-1} = \boxed{\frac{x+\sqrt{x}}{x-1}}$

(10) $\frac{\sqrt{x}+\sqrt{3}}{\sqrt{x}-\sqrt{2}} = \frac{\sqrt{x}+\sqrt{3}}{\sqrt{x}-\sqrt{2}}\left(\frac{\sqrt{x}+\sqrt{2}}{\sqrt{x}+\sqrt{2}}\right) = \frac{\sqrt{x}\sqrt{x}+\sqrt{x}\sqrt{2}+\sqrt{3}\sqrt{x}+\sqrt{3}\sqrt{2}}{\sqrt{x}\sqrt{x}+\sqrt{x}\sqrt{2}-\sqrt{2}\sqrt{x}-\sqrt{2}\sqrt{2}} = \boxed{\frac{x+\sqrt{3x}+\sqrt{2x}+\sqrt{6}}{x-2}}$

(11) $\frac{x+3}{3+2\sqrt{x}} = \frac{x+3}{3+2\sqrt{x}}\left(\frac{3-2\sqrt{x}}{3-2\sqrt{x}}\right) = \frac{3x-2x\sqrt{x}+9-6\sqrt{x}}{3(3)-6\sqrt{x}+6\sqrt{x}-4\sqrt{x}\sqrt{x}} = \boxed{\frac{-2x\sqrt{x}+3x-6\sqrt{x}+9}{-4x+9}}$

2.6 Problems

(1) $125^{1/3} = \sqrt[3]{125} = \boxed{5}$ because $5^3 = (5)(5)(5) = 125$

(2) $36^{1/2} = \sqrt{36} = \boxed{6}$

(3) $(-32)^{1/5} = \sqrt[5]{-32} = \boxed{-2}$ because $(-2)^5 = (-2)(-2)(-2)(-2)(-2) = -32$

(4) $(256x^{12})^{1/4} = 256^{1/4}(x^{12})^{1/4} = \sqrt[4]{256}\,x^{12/4} = \boxed{4x^3}$

(5) $16^{3/2} = \left(16^{1/2}\right)^3 = \left(\sqrt{16}\right)^3 = 4^3 = \boxed{64}$

(6) $27^{2/3} = \left(27^{1/3}\right)^2 = \left(\sqrt[3]{27}\right)^2 = 3^2 = \boxed{9}$

(7) $81^{3/4} = \left(81^{1/4}\right)^3 = \left(\sqrt[4]{81}\right)^3 = 3^3 = \boxed{27}$

(8) $(-125)^{5/3} = \left[(-125)^{1/3}\right]^5 = \left(\sqrt[3]{-125}\right)^5 = (-5)^5 = \boxed{-3125}$

(9) $(4x^6)^{5/2} = 4^{5/2}(x^6)^{5/2} = \left(4^{1/2}\right)^5\left[(x^6)^{1/2}\right]^5 = \left(\sqrt{4}\right)^5\left(x^{6/2}\right)^5 = 2^5(x^3)^5 = \boxed{32x^{15}}$

(10) $(32x^5)^{2/5} = 32^{2/5}(x^5)^{2/5} = \left(32^{1/5}\right)^2\left[(x^5)^{1/5}\right]^2 = \left(\sqrt[5]{32}\right)^2(x^1)^2 = 2^2x^2 = \boxed{4x^2}$

(11) $10^{-2} = \frac{1}{10^2} = \boxed{\frac{1}{100}}$ Alternate answer: 0.01

(12) $\left(4x^{2/3}\right)^{-3} = 4^{-3}\left(x^{2/3}\right)^{-3} = \frac{1}{4^3}\frac{1}{(x^{2/3})^3} = \boxed{\frac{1}{64x^2}}$ Note: $(x^a)^b = x^{ab}$.

(13) $256^{-1/4} = \dfrac{1}{256^{1/4}} = \dfrac{1}{\sqrt[4]{256}} = \boxed{\dfrac{1}{4}}$

(14) $(-27x)^{-1/3} = \dfrac{1}{(-27x)^{1/3}} = \dfrac{1}{\sqrt[3]{-27}\,x^{1/3}} = \boxed{-\dfrac{1}{3x^{1/3}}}$ Alternate answer: $-\dfrac{x^{-1/3}}{3}$

(15) $\left(\dfrac{125}{64}\right)^{-2/3} = \left(\dfrac{64}{125}\right)^{2/3} = \left(\dfrac{64^{1/3}}{125^{1/3}}\right)^{2} = \left(\dfrac{\sqrt[3]{64}}{\sqrt[3]{125}}\right)^{2} = \left(\dfrac{4}{5}\right)^{2} = \boxed{\dfrac{16}{25}}$ Note: $\left(\dfrac{x}{y}\right)^{a} = \dfrac{x^{a}}{y^{a}}$.

(16) $\left(\dfrac{16x^{6}}{81}\right)^{-3/4} = \left(\dfrac{81}{16x^{6}}\right)^{3/4} = \left[\dfrac{81^{1/4}}{16^{1/4}(x^{6})^{1/4}}\right]^{3} = \left(\dfrac{\sqrt[4]{81}}{\sqrt[4]{16}\,x^{6/4}}\right)^{3} = \left(\dfrac{3}{2x^{3/2}}\right)^{3} = \boxed{\dfrac{27}{8x^{9/2}}} = \dfrac{27}{8x^{4}\sqrt{x}} = \boxed{\dfrac{27\sqrt{x}}{8x^{5}}}$

(17) $\left(\dfrac{1}{64}\right)^{-1/2} = 64^{1/2} = \sqrt{64} = \boxed{8}$

(18) $\left(\dfrac{1}{81x^{8}}\right)^{-1/4} = (81x^{8})^{1/4} = 81^{1/4}(x^{8})^{1/4} = \sqrt[4]{81}\,x^{8/4} = \boxed{3x^{2}}$

(19) $\left(\dfrac{27x^{9/2}}{64}\right)^{-4/3} = \left(\dfrac{64}{27x^{9/2}}\right)^{4/3} = \left[\dfrac{64^{1/3}}{27^{1/3}(x^{9/2})^{1/3}}\right]^{4} = \left(\dfrac{\sqrt[3]{64}}{\sqrt[3]{27}\,x^{3/2}}\right)^{4} = \left(\dfrac{4}{3x^{3/2}}\right)^{4} = \boxed{\dfrac{256}{81x^{6}}}$

Notes: $\left(x^{9/2}\right)^{1/3} = x^{(9\cdot1)/(2\cdot3)} = x^{9/6} = x^{3/2}$ and $\left(x^{3/2}\right)^{4} = x^{(3\cdot4)/2} = x^{12/2} = x^{6}$ according to the rule $(x^{a})^{b} = x^{ab}$. Also, $\dfrac{9}{2}\cdot\dfrac{1}{3} = \dfrac{9}{6} = \dfrac{3}{2}$.

(20) $\left(\dfrac{32x^{-10}}{243}\right)^{-2/5} = \left(\dfrac{243}{32x^{-10}}\right)^{2/5} = \left(\dfrac{243x^{10}}{32}\right)^{2/5} = \left[\dfrac{243^{1/5}(x^{10})^{1/5}}{32^{1/5}}\right]^{2} = \left(\dfrac{3x^{2}}{2}\right)^{2} = \boxed{\dfrac{9x^{4}}{4}}$

2.7 Problems

(1) $6x\,\sqrt{3x^{3}+5x} = \sqrt{(6x)^{2}(3x^{3}+5x)} = \sqrt{36x^{2}(3x^{3}+5x)} = \boxed{\sqrt{108x^{5}+180x^{3}}}$

(2) $5x^{2}\,\sqrt{7x-6} = \sqrt{(5x^{2})^{2}(7x-6)} = \sqrt{25x^{4}(7x-6)} = \boxed{\sqrt{175x^{5}-150x^{4}}}$

(3) $3x\,\sqrt[3]{x^{4}+2x^{2}+5} = \sqrt[3]{(3x)^{3}(x^{4}+2x^{2}+5)} = \sqrt[3]{27x^{3}(x^{4}+2x^{2}+5)} =$
$\boxed{\sqrt[3]{27x^{7}+54x^{5}+135x^{3}}}$

(4) $2x^{2}\,\sqrt[5]{5x-9} = \sqrt[5]{(2x^{2})^{5}(5x-9)} = \sqrt[5]{32x^{10}(5x-9)} = \boxed{\sqrt[5]{160x^{11}-288x^{10}}}$

(5) $8x^{3}\,\sqrt{5x^{7}-3x^{4}+x} = \sqrt{(8x^{3})^{2}(5x^{7}-3x^{4}+x)} = \sqrt{64x^{6}(5x^{7}-3x^{4}+x)} =$
$\boxed{\sqrt{320x^{13}-192x^{10}+64x^{7}}}$

(6) $7x^{4}\,\sqrt[3]{2x^{3}+7x} = \sqrt[3]{(7x^{4})^{3}(2x^{3}+7x)} = \sqrt[3]{343x^{12}(2x^{3}+7x)} = \boxed{\sqrt[3]{686x^{15}+2401x^{13}}}$

(7) $2x\,\sqrt[4]{x^{4}-8x^{2}-4} = \sqrt[4]{(2x)^{4}(x^{4}-8x^{2}-4)} = \sqrt[4]{16x^{4}(x^{4}-8x^{2}-4)} =$
$\boxed{\sqrt[4]{16x^{8}-128x^{6}-64x^{4}}}$

(8) $10x^3 \sqrt{15x^5 - 6x^3 + 9x} = \sqrt{(10x^3)^2(15x^5 - 6x^3 + 9x)} =$

$\sqrt{100x^6(15x^5 - 6x^3 + 9x)} = \boxed{\sqrt{1500x^{11} - 600x^9 + 900x^7}}$

(9) $3x(x^2 + 4x)^{1/3} = [(3x)^3(x^2 + 4x)]^{1/3} = [27x^3(x^2 + 4x)]^{1/3} = \boxed{(27x^5 + 108x^4)^{1/3}}$

(10) $5x^2(3x - 2)^{1/4} = [(5x^2)^4(3x - 2)]^{1/4} = [625x^8(3x - 2)]^{1/4} =$

$\boxed{(1875x^9 - 1250x^8)^{1/4}}$

(11) $4x^4(5x^2 + 6)^{2/3} = \left[(4x^4)^{3/2}(5x^2 + 6)\right]^{2/3} = \left[4^{3/2}x^{12/2}(5x^2 + 6)\right]^{2/3} =$

$\left[(\sqrt{4})^3 x^6(5x^2 + 6)\right]^{2/3} = [8x^6(5x^2 + 6)]^{2/3} = \boxed{(40x^8 + 48x^6)^{2/3}}$

(12) $32x^5(4x^3 - 8x)^{5/4} = \left[(32x^5)^{4/5}(4x^3 - 8x)\right]^{5/4} = \left[32^{4/5}x^{20/5}(4x^3 - 8x)\right]^{5/4} =$

$\left[(\sqrt[5]{32})^4 x^4(4x^3 - 8x)\right]^{5/4} = [16x^4(4x^3 - 8x)]^{5/4} = \boxed{(64x^7 - 128x^5)^{5/4}}$

(13) $64x^{12}(x^2 + 3x - 7)^{3/5} = \left[(64x^{12})^{5/3}(x^2 + 3x - 7)\right]^{3/5} = \left[64^{5/3}x^{60/3}(x^2 + 3x - 7)\right]^{3/5}$

$= \left[(\sqrt[3]{64})^5 x^{20}(x^2 + 3x - 7)\right]^{3/5} = [1024x^{20}(x^2 + 3x - 7)]^{3/5} =$

$\boxed{(1024x^{22} + 3072x^{21} - 7168x^{20})^{3/5}}$

(14) $125x^6(9x^3 - 6)^{3/2} = \left[(125x^6)^{2/3}(9x^3 - 6)\right]^{3/2} = \left[125^{2/3}x^{12/3}(9x^3 - 6)\right]^{3/2} =$

$\left[(\sqrt[3]{125})^2 x^4(9x^3 - 6)\right]^{3/2} = [25x^4(9x^3 - 6)]^{3/2} = \boxed{(225x^7 - 150x^4)^{3/2}}$

(15) $256x^{24}(5x^4 + 8x)^{4/3} = \left[(256x^{24})^{3/4}(5x^4 + 8x)\right]^{4/3} = \left[256^{3/4}x^{72/4}(5x^4 + 8x)\right]^{4/3}$

$= \left[(\sqrt[4]{256})^3 x^{18}(5x^4 + 8x)\right]^{4/3} = [64x^{18}(5x^4 + 8x)]^{4/3} = \boxed{(320x^{22} + 512x^{19})^{4/3}}$

(16) $243x^{10}(2x^8 - 3x^6 + 6x^4)^{-5/2} = \left[(243x^{10})^{-2/5}(2x^8 - 3x^6 + 6x^4)\right]^{-5/2} =$

$\left[243^{-2/5}x^{-20/5}(2x^8 - 3x^6 + 6x^4)\right]^{-5/2} = \left[(\sqrt[5]{243})^{-2} x^{-4}(2x^8 - 3x^6 + 6x^4)\right]^{-5/2} =$

$[3^{-2}x^{-4}(2x^8 - 3x^6 + 6x^4)]^{-5/2} = \left[\frac{1}{3^2 x^4}(2x^8 - 3x^6 + 6x^4)\right]^{-5/2}$

$= \left[\frac{1}{9x^4}(2x^8 - 3x^6 + 6x^4)\right]^{-5/2} = \frac{2x^8}{9x^4} - \frac{3x^6}{9x^4} + \frac{6x^4}{9x^4} = \boxed{\left(\frac{2}{9}x^4 - \frac{x^2}{3} + \frac{2}{3}\right)^{-5/2}}$

2.8 Problems

(1) $9 + 2\sqrt{3} - 4 + 7\sqrt{3} = (9 - 4) + (2 + 7)\sqrt{3} = \boxed{5 + 9\sqrt{3}}$

(2) $6\sqrt{5} + 5\sqrt{6} - 2\sqrt{5} + 9\sqrt{6} = (6 - 2)\sqrt{5} + (5 + 9)\sqrt{6} = \boxed{4\sqrt{5} + 14\sqrt{6}}$

(3) $x\sqrt{2} + 4\sqrt{x} + 8\sqrt{x} - 3x\sqrt{2} = (x - 3x)\sqrt{2} + (4 + 8)\sqrt{x} = \boxed{-2x\sqrt{2} + 12\sqrt{x}}$

(4) $x^{3/2} + x - x\sqrt{x} = x\sqrt{x} + x - x\sqrt{x} = \boxed{x}$ Note: $x^{3/2} = x^1 x^{1/2} = x\sqrt{x}$.

(5) $5\sqrt{2} + \frac{6}{\sqrt{2}} = 5\sqrt{2} + \frac{6}{\sqrt{2}}\frac{\sqrt{2}}{\sqrt{2}} = 5\sqrt{2} + \frac{6\sqrt{2}}{2} = 5\sqrt{2} + 3\sqrt{2} = \boxed{8\sqrt{2}}$

(6) $\frac{7\sqrt{x}}{x} - \frac{4}{\sqrt{x}} = \frac{7\sqrt{x}}{x} - \frac{4}{\sqrt{x}}\frac{\sqrt{x}}{\sqrt{x}} = \frac{7\sqrt{x}}{x} - \frac{4\sqrt{x}}{x} = \boxed{\frac{3\sqrt{x}}{x}}$ Note: Although this answer is equivalent to $\frac{3}{\sqrt{x}}$, the answer $\frac{3}{\sqrt{x}}$ is not in standard form.

(7) $(9 + \sqrt{5})(3 - \sqrt{5}) = 9(3) - 9\sqrt{5} + 3\sqrt{5} - \sqrt{5}\sqrt{5} = 27 - 6\sqrt{5} - 5 = \boxed{22 - 6\sqrt{5}}$

(8) $(\sqrt{3} - \sqrt{2})(\sqrt{3} - \sqrt{2}) = \sqrt{3}\sqrt{3} - \sqrt{3}\sqrt{2} - \sqrt{2}\sqrt{3} + \sqrt{2}\sqrt{2} = 3 - \sqrt{6} - \sqrt{6} + 2 = \boxed{5 - 2\sqrt{6}}$

(9) $(x + \sqrt{3})(x - \sqrt{6}) = x^2 - x\sqrt{6} + x\sqrt{3} - \sqrt{3}\sqrt{6} = x^2 - x\sqrt{6} + x\sqrt{3} - \sqrt{3}\sqrt{2}\sqrt{3} = $
$\boxed{x^2 - x\sqrt{6} + x\sqrt{3} - 3\sqrt{2}}$ Note: $\sqrt{3}\sqrt{6} = \sqrt{3}\sqrt{(2)(3)} = \sqrt{3}\sqrt{2}\sqrt{3} = 3\sqrt{2}$.

(10) $(x + \sqrt{x})(x + \sqrt{x}) = x^2 + x\sqrt{x} + x\sqrt{x} + \sqrt{x}\sqrt{x} = \boxed{x^2 + 2x\sqrt{x} + x}$

(11) $\sqrt{x}(x\sqrt{x} - 4x + 3\sqrt{x} - 8) = x\sqrt{x}\sqrt{x} - 4x\sqrt{x} + 3\sqrt{x}\sqrt{x} - 8\sqrt{x} = $
$\boxed{x^2 - 4x\sqrt{x} + 3x - 8\sqrt{x}}$

(12) $(x - 2\sqrt{x} + \sqrt{3})(\sqrt{x} - \sqrt{3}) = x\sqrt{x} - x\sqrt{3} - 2\sqrt{x}\sqrt{x} + 2\sqrt{x}\sqrt{3} + \sqrt{3}\sqrt{x} - \sqrt{3}\sqrt{3}$
$= x\sqrt{x} - x\sqrt{3} - 2x + 2\sqrt{3x} + \sqrt{3x} - 3 = \boxed{x\sqrt{x} - 2x - x\sqrt{3} + 3\sqrt{3x} - 3}$

(13) $(x - \sqrt{x})^3 = (x - \sqrt{x})(x - \sqrt{x})(x - \sqrt{x}) = (x - \sqrt{x})(x^2 - 2x\sqrt{x} + x)$
$= x^3 - 2x^2\sqrt{x} + x^2 - x^2\sqrt{x} + 2x^2 - x\sqrt{x} = \boxed{x^3 - 3x^2\sqrt{x} + 3x^2 - x\sqrt{x}}$
Note: Some of these answers can be factored. For example, this answer can be expressed as $x\sqrt{x}(x\sqrt{x} - 3x + 3\sqrt{x} - 1)$.

(14) $(3x + \sqrt{2})^3 = (3x + \sqrt{2})(3x + \sqrt{2})(3x + \sqrt{2}) = (3x + \sqrt{2})(9x^2 + 6x\sqrt{2} + 2)$
$= 27x^3 + 18x^2\sqrt{2} + 6x + 9x^2\sqrt{2} + 6x(2) + 2\sqrt{2} = \boxed{27x^3 + 27x^2\sqrt{2} + 18x + 2\sqrt{2}}$

(15) $\sqrt{2x^9}\sqrt{20x^5} = \sqrt{2x^9 20x^5} = \sqrt{40x^{14}} = x^7\sqrt{40} = x^7\sqrt{4}\sqrt{10} = \boxed{2x^7\sqrt{10}}$

(16) $\sqrt{6x^7}\sqrt{10x^5}\sqrt{15x^3} = \sqrt{6x^7 10x^5 15x^3} = \sqrt{900x^{15}} = \boxed{30x^7\sqrt{x}}$ Recall Sec. 2.2.

(17) $\frac{\sqrt{21x^7}}{\sqrt{7x}} = \sqrt{\frac{21x^7}{7x}} = \sqrt{3x^6} = \boxed{x^3\sqrt{3}}$

(18) $\frac{\sqrt{30x^{21}}}{\sqrt{6x^{13}}} = \sqrt{\frac{30x^{21}}{6x^{13}}} = \sqrt{5x^8} = \boxed{x^4\sqrt{5}}$

2.9 Problems

(1) $x = \boxed{\pm\sqrt{7}}$ Check: $6 + 5(\pm\sqrt{7})^2 = 6 + 5(7) = 41$.

(2) $x = \boxed{\pm3\sqrt{2}}$ Check: $7(\pm3\sqrt{2})^2 - 24 = 7(9)(2) - 24 = 126 - 24 = 102$ agrees

with $4x^2 + 30 = 4(\pm3\sqrt{2})^2 + 30 = 4(9)(2) + 30 = 72 + 30 = 102$.

Notes: $3\sqrt{2}$ is equivalent to $\sqrt{18}$ since $\sqrt{18} = \sqrt{9(2)} = 3\sqrt{2}$, but $\sqrt{18}$ isn't in standard

form. Observe that $(\pm3\sqrt{2})^2 = 3^2(\sqrt{2})^2 = 9(2) = 18$.

(3) $x = \boxed{\sqrt[3]{5}}$ Check: $10 - (\sqrt[3]{5})^3 = 10 - 5 = 5$ agrees with $(\sqrt[3]{5})^3 = 5$.

(4) $x = \boxed{-\sqrt[5]{11}}$ Check: $2(-\sqrt[5]{11})^5 - 59 = 2(-11) - 59 = -22 - 59 = -81$ agrees

with $6(-\sqrt[5]{11})^5 - 15 = 6(-11) - 15 = -66 - 15 = -81$.

(5) $x = \boxed{\pm2}$ Check: $9(\pm2)^4 + 14 = 9(16) + 14 = 144 + 14 = 158$ agrees with

$6(\pm2)^4 + 62 = 6(16) + 62 = 96 + 62 = 158$.

(6) $x = \boxed{\pm2\sqrt{5}}$ Check: $140 - (\pm2\sqrt{5})^2 = 140 - 4(5) = 140 - 20 = 120$ agrees with

$20 + 5(\pm2\sqrt{5})^2 = 20 + 5(4)(5) = 20 + 100 = 120$.

Notes: $2\sqrt{5}$ is equivalent to $\sqrt{20}$ since $\sqrt{20} = \sqrt{4(5)} = 2\sqrt{5}$, but $\sqrt{20}$ isn't in standard

form. Observe that $(\pm2\sqrt{5})^2 = 2^2(\sqrt{5})^2 = 4(5) = 20$.

(7) $x = \boxed{\sqrt[3]{12}}$ Check: $\frac{8}{(\sqrt[3]{12})^3} = \frac{8}{12} = \frac{2}{3}$. Note: The answer $\sqrt[3]{12}$ can't be simplified

further because 12 is not a multiple of a perfect **cube** like 8, 27, or 64.

(8) $x = \boxed{\pm\sqrt{6}}$ Check: $\frac{2}{\pm\sqrt{6}} = \pm\frac{2}{\sqrt{6}}\frac{\sqrt{6}}{\sqrt{6}} = \pm\frac{2\sqrt{6}}{6} = \pm\frac{\sqrt{6}}{3}$ agrees with $\frac{\pm\sqrt{6}}{3}$.

(9) $x = \boxed{\pm2\sqrt{3}}$ Check: $\frac{1}{x^2} + \frac{1}{4} = \frac{1}{(\pm2\sqrt{3})^2} + \frac{1}{4} = \frac{1}{4(3)} + \frac{1}{4} = \frac{1}{12} + \frac{1}{4} = \frac{1}{12} + \frac{3}{12} = \frac{4}{12} = \frac{1}{3}$.

Note: $2\sqrt{3}$ is equivalent to $\sqrt{12}$ since $\sqrt{12} = \sqrt{4(3)} = 2\sqrt{3}$, but $\sqrt{20}$ isn't in standard form.

(10) $x = \boxed{-\sqrt[3]{3}}$ Check: $\frac{1}{6} - \frac{1}{\left(-\sqrt[3]{3}\right)^3} = \frac{1}{6} - \frac{1}{-3} = \frac{1}{6} + \frac{1}{3} = \frac{1}{6} + \frac{2}{6} = \frac{3}{6} = \frac{1}{2}$.

2.10 Problems

(1) $x = \boxed{24}$ Check: $\sqrt{6(24)} - 3 = \sqrt{144} - 3 = 12 - 3 = 9$.

(2) $x = \boxed{25}$ Check: $10 - \sqrt{25} = 10 - 5 = 5$ agrees with $\sqrt{25} = 5$.

(3) $x = \boxed{39}$ Check: $14 + \sqrt{39 - 3} = 14 + \sqrt{36} = 14 + 6 = 20$.

(4) $x = \boxed{32}$ Check: $7\sqrt{2(32)} + 5 = 7\sqrt{64} + 5 = 7(8) + 5 = 56 + 5 = 61$ agrees with

$37 + 3\sqrt{2(32)} = 37 + 3\sqrt{64} = 37 + 3(8) = 37 + 24 = 61$.

(5) $x = \boxed{57}$ Check: $\sqrt{2(57) + 7} - 4 = \sqrt{114 + 7} - 4 = \sqrt{121} - 4 = 11 - 4 = 7$.

(6) $x = \boxed{\pm 3}$ Check: $21 - \sqrt{45 - (\pm 3)^2} = 21 - \sqrt{45 - 9} = 21 - \sqrt{36} = 21 - 6 = 15$.

(7) $x = \boxed{\frac{25}{4}}$ Check: $\frac{6}{\sqrt{25/4}} = \frac{6}{5/2} = 6\left(\frac{2}{5}\right) = \frac{12}{5}$.

(8) $x = \boxed{18}$ Check: $\frac{1}{\sqrt{2(18)}} - \frac{1}{18} = \frac{1}{\sqrt{36}} - \frac{1}{18} = \frac{1}{6} - \frac{1}{18} = \frac{3}{18} - \frac{1}{18} = \frac{2}{18} = \frac{1}{9}$.

(9) $x = \boxed{\frac{25}{16}}$ Check: $4 - \frac{2}{\sqrt{25/16}} = 4 - \frac{2}{5/4} = 4 - 2\left(\frac{4}{5}\right) = 4 - \frac{8}{5} = \frac{20}{5} - \frac{8}{5} = \frac{12}{5}$ agrees

with $\frac{3}{\sqrt{x}} = \frac{3}{\sqrt{25/16}} = \frac{3}{5/4} = 3\left(\frac{4}{5}\right) = \frac{12}{5}$.

(10) $x = \boxed{\frac{4}{3}}$ Check: $\sqrt{\frac{4}{3}} - \frac{1}{\sqrt{3}} = \frac{\sqrt{4}}{\sqrt{3}} - \frac{1}{\sqrt{3}} = \frac{2}{\sqrt{3}} - \frac{1}{\sqrt{3}} = \frac{1}{\sqrt{3}}$.

(11) $x = \boxed{279}$ Check: $\sqrt{279 + 45} = \sqrt{324} = 18$ agrees with $2 + \sqrt{279 - 23} = 2 + \sqrt{256} = 2 + 16 = 18$.

(12) $x = \boxed{48}$ Check: $\sqrt{3(48) + 81} - 3 = \sqrt{144 + 81} - 3 = \sqrt{225} - 3 = 15 - 3 = 12$

agrees with $\sqrt{3(48)} = \sqrt{144} = 12$.

(13) $x = \boxed{5 + 2\sqrt{6}}$ Calculator check: $\sqrt{3} \approx 1.732$ agrees with $\sqrt{5 + 2\sqrt{6}} - \sqrt{2} \approx$

$\sqrt{5 + 2(2.449)} - 1.414 \approx \sqrt{5 + 4.898} - 1.414 \approx \sqrt{9.898} - 1.414 \approx 3.146 - 1.414 \approx 1.732$.

(14) $x = \boxed{\frac{5}{2}} = \boxed{2.5}$ Check: $10 - \sqrt{2\left(\frac{5}{2}\right) + 11} = 10 - \sqrt{5 + 11} = 10 - \sqrt{16} = 10 - 4 = 6$

agrees with $\sqrt{2\left(\frac{5}{2}\right) + 31} = \sqrt{5 + 31} = \sqrt{36} = 6$.

2.11 Problems

(1) $x = \boxed{256}$ Check: $16 + 256^{3/4} = 16 + \left(256^{1/4}\right)^3 = 16 + \left(\sqrt[4]{256}\right)^3 = 16 + 4^3 = 16 + 64 = 80.$

(2) $x = \boxed{\pm 32}$ Check: $8(\pm 32)^{2/5} - 12 = 8\left[(\pm 32)^{1/5}\right]^2 - 12 = 8\left(\sqrt[5]{\pm 32}\right)^2 - 12 = 8(\pm 2)^2 - 12 = 8(4) - 12 = 32 - 12 = 20$ agrees with $5(\pm 32)^{2/5} = 5\left[(\pm 32)^{1/5}\right]^2 = 5\left(\sqrt[5]{\pm 32}\right)^2 = 5(\pm 2)^2 = 5(4) = 20.$

(3) $x = \boxed{-27}$ Check: $(-27)^{5/3} + 500 = \left[(-27)^{1/3}\right]^5 + 500 = \left(\sqrt[3]{-27}\right)^5 + 500 = (-3)^5 + 500 = -243 + 500 = 257$ agrees with $14 - (-27)^{5/3} = 14 - \left[(-27)^{1/3}\right]^5 = 14 - \left(\sqrt[3]{-27}\right)^5 = 14 - (-3)^5 = 14 - (-243) = 14 + 243 = 257.$

(4) $x = \boxed{9}$ Check: $3(9)^{3/2} - 35 = 3\left(9^{1/2}\right)^3 - 35 = 3\left(\sqrt{9}\right)^3 - 35 = 3(3)^3 - 35 = 3(27) - 35 = 81 - 35 = 46$ agrees with $100 - 2x^{3/2} = 100 - 2(9)^{3/2} = 100 - 2\left(9^{1/2}\right)^3 = 100 - 2\left(\sqrt{9}\right)^3 = 100 - 2(3)^3 = 100 - 2(27) = 100 - 54 = 46.$

(5) $x = \boxed{\pm 343}$ Check: $6(\pm 343)^{2/3} - 48 = 6\left[(\pm 343)^{1/3}\right]^2 - 48 = 6\left(\sqrt[3]{\pm 343}\right)^2 - 48 = 6(\pm 7)^2 - 48 = 6(49) - 48 = 294 - 48 = 246$ agrees with $4(\pm 343)^{2/3} + 50 = 4\left[(\pm 343)^{1/3}\right]^2 + 50 = 4\left(\sqrt[3]{\pm 343}\right)^2 + 50 = 4(\pm 7)^2 + 50 = 4(49) + 50 = 196 + 50 = 246.$

(6) $x = \boxed{36}$ Check: $36\sqrt{36} - 56 = 36(6) - 56 = 216 - 56 = 160.$

Note: $x\sqrt{x} = x^1 x^{1/2} = x^{3/2}.$

(7) $x = \boxed{\pm 27}$ Check: $\dfrac{54}{(\pm 27)^{4/3}} = \dfrac{54}{\left[(\pm 27)^{1/3}\right]^4} = \dfrac{54}{\left(\sqrt[3]{\pm 27}\right)^4} = \dfrac{54}{(\pm 3)^4} = \dfrac{54}{81} = \dfrac{54 \div 27}{81 \div 27} = \dfrac{2}{3}.$

(8) $x = \boxed{64}$ Check: $\dfrac{8}{64^{1/2}} = \dfrac{8}{\sqrt{64}} = \dfrac{8}{8} = 1$ agrees with $\dfrac{64^{1/3}}{4} = \dfrac{4}{4} = 1.$

(9) $x = \boxed{\pm 216}$ Check: $\dfrac{1}{(\pm 216)^{2/3}} + \dfrac{1}{18} = \dfrac{1}{\left[(\pm 216)^{1/3}\right]^2} + \dfrac{1}{18} = \dfrac{1}{(\pm 6)^2} + \dfrac{1}{18} = \dfrac{1}{36} + \dfrac{2}{36} = \dfrac{3}{36} = \dfrac{1}{12}.$

(10) $x = \boxed{16}$ Check: $\dfrac{5}{12} - \dfrac{16}{16^{3/2}} = \dfrac{5}{12} - \dfrac{16}{\left(16^{1/2}\right)^3} = \dfrac{5}{12} - \dfrac{16}{\left(\sqrt{16}\right)^3} = \dfrac{5}{12} - \dfrac{16}{4^3} = \dfrac{5}{12} - \dfrac{16}{64} = \dfrac{5}{12} - \dfrac{1}{4} = \dfrac{5}{12} - \dfrac{3}{12} = \dfrac{2}{12} = \dfrac{1}{6}.$

3 Complex Numbers

3.1 Problems

(1) $\sqrt{-25} = \sqrt{25(-1)} = \sqrt{25}\sqrt{-1} = \boxed{5i}$

(2) $\sqrt{-17} = \sqrt{17(-1)} = \sqrt{17}\sqrt{-1} = \boxed{i\sqrt{17}}$

(3) $\sqrt{-49} = \sqrt{49(-1)} = \sqrt{49}\sqrt{-1} = \boxed{7i}$

(4) $\sqrt{-23} = \sqrt{23(-1)} = \sqrt{23}\sqrt{-1} = \boxed{i\sqrt{23}}$

(5) $\sqrt{-8} = \sqrt{4(2)(-1)} = \sqrt{4}\sqrt{2}\sqrt{-1} = \boxed{2i\sqrt{2}}$

(6) $\sqrt{-81} = \sqrt{81(-1)} = \sqrt{81}\sqrt{-1} = \boxed{9i}$

(7) $\sqrt{-15} = \sqrt{15(-1)} = \sqrt{15}\sqrt{-1} = \boxed{i\sqrt{15}}$

Note: Although $15 = 5(3)$, neither 5 nor 3 is a perfect square.

(8) $\sqrt{-48} = \sqrt{16(3)(-1)} = \sqrt{16}\sqrt{3}\sqrt{-1} = \boxed{4i\sqrt{3}}$

(9) $\sqrt{-144} = \sqrt{144(-1)} = \sqrt{144}\sqrt{-1} = \boxed{12i}$

(10) $\sqrt{-63} = \sqrt{9(7)(-1)} = \sqrt{9}\sqrt{7}\sqrt{-1} = \boxed{3i\sqrt{7}}$

(11) $\sqrt{-6} = \sqrt{(-1)} = \sqrt{6}\sqrt{-1} = \boxed{i\sqrt{6}}$

(12) $\sqrt{-75} = \sqrt{25(3)(-1)} = \sqrt{25}\sqrt{3}\sqrt{-1} = \boxed{5i\sqrt{3}}$

(13) $\sqrt{-64} = \sqrt{64(-1)} = \sqrt{64}\sqrt{-1} = \boxed{8i}$

(14) $\sqrt{-42} = \sqrt{42(-1)} = \sqrt{42}\sqrt{-1} = \boxed{i\sqrt{42}}$

Note: Although $42 = 7(3)(2)$, neither 7, 3, nor 2 is a perfect square.

(15) $\sqrt{-24} = \sqrt{4(6)(-1)} = \sqrt{4}\sqrt{6}\sqrt{-1} = \boxed{2i\sqrt{6}}$

(16) $\sqrt{-256} = \sqrt{256(-1)} = \sqrt{256}\sqrt{-1} = \boxed{16i}$

(17) $\sqrt{-72} = \sqrt{36(2)(-1)} = \sqrt{36}\sqrt{2}\sqrt{-1} = \boxed{6i\sqrt{2}}$

(18) $\sqrt{-245} = \sqrt{49(5)(-1)} = \sqrt{49}\sqrt{5}\sqrt{-1} = \boxed{7i\sqrt{5}}$

(19) $\sqrt{-53} = \sqrt{53(-1)} = \sqrt{53}\sqrt{-1} = \boxed{i\sqrt{53}}$

(20) $\sqrt{-361} = \sqrt{361(-1)} = \sqrt{361}\sqrt{-1} = \boxed{19i}$

3.2 Problems

(1) Given $z = 4 + 5i$, $\text{Re}(z) = 4$ and $\text{Im}(z) = 5$.

(2) Given $w = 2 + i$, $\text{Re}(w) = 2$ and $\text{Im}(w) = 1$.

(3) Given $u = -7 + 6i$, $\text{Re}(u) = -7$ and $\text{Im}(u) = 6$.

(4) Given $z = -3 - 3i$, $\text{Re}(z) = -3$ and $\text{Im}(z) = -3$.

(5) Given $v = 8i$, $\text{Re}(v) = 0$ and $\text{Im}(v) = 8$. This number is purely imaginary.

(6) Given $t = 5 - i$, $\text{Re}(t) = 5$ and $\text{Im}(t) = -1$.

(7) Given $w = -10 + 2i$, $\text{Re}(w) = -10$ and $\text{Im}(w) = 2$.

(8) Given $z = 2 - \frac{i}{2}$, $\text{Re}(z) = 2$ and $\text{Im}(z) = -\frac{1}{2}$.

(9) Given $d = 12{,}073$, $\text{Re}(d) = 12{,}073$ and $\text{Im}(d) = 0$. This number is purely real.

(10) Given $q = 0.4 + i\sqrt{2}$, $\text{Re}(q) = 0.4$ or $\frac{2}{5}$ and $\text{Im}(q) = \sqrt{2}$.

3.3 Problems

(1) $(5 + 9i) + (6 + 4i) = (5 + 6) + (9 + 4)i = \boxed{11 + 13i}$

(2) $(6 - 2i) + (4 + 7i) = (6 + 4) + (-2 + 7)i = \boxed{10 + 5i}$

(3) $(14 + 7i) + (16 - 3i) = (14 + 16) + [7 + (-3)]i = \boxed{30 + 4i}$

(4) $(11 - 3i) + (6 - 6i) = (11 + 6) + [-3 + (-6)]i = \boxed{17 - 9i}$

(5) $(15 - 8i) + (-2 + i) = [15 + (-2)] + (-8 + 1)i = \boxed{13 - 7i}$

(6) $(8 + 6i) - (9 + 3i) = (8 - 9) + (6 - 3)i = \boxed{-1 + 3i}$

(7) $(4 + 7i) - (10 - 2i) = (4 - 10) + [7 - (-2)]i = -6 + (7 + 2)i = \boxed{-6 + 9i}$

(8) $(12 - 5i) - (3 + 4i) = (12 - 3) + (-5 - 4)i = \boxed{9 - 9i}$

(9) $(16 - 12i) - (9 - 7i) = (16 - 9) + [-12 - (-7)]i = 7 + (-12 + 7)i = \boxed{7 - 5i}$

(10) $(11 + 14i) - (-8 + 6i) = [11 - (-8)] + (14 - 6)i = (11 + 8) + 8i = \boxed{19 + 8i}$

3.4 Problems

(1) $(6 + 3i)(8 + 4i) = 6(8) + 6(4i) + 3i(8) + 3i(4i) = 48 + 24i + 24i + 12(-1) =$
$48 + 48i - 12 = \boxed{36 + 48i}$

(2) $(3+i)(9-i) = 3(9) + 3(-i) + i(9) + i(-i) = 27 - 3i + 9i - (-1) =$
$27 + 6i + 1 = \boxed{28 + 6i}$

(3) $(5-6i)(3+2i) = 5(3) + 5(2i) - 6i(3) - 6i(2i) = 15 + 10i - 18i - 12(-1) =$
$15 - 8i + 12 = \boxed{27 - 8i}$

(4) $(8-5i)(9-8i) = 8(9) + 8(-8i) - 5i(9) - 5i(-8i) = 72 - 64i - 45i + 40(-1) =$
$72 - 109i - 40 = \boxed{32 - 109i}$

(5) $(-5+9i)(7-3i) = -5(7) - 5(-3i) + 9i(7) + 9i(-3i) =$
$-35 + 15i + 63i - 27(-1) = -35 + 78i + 27 = \boxed{-8 + 78i}$

(6) $-7i(4-i) = -7i(4) - 7i(-i) = -28i + 7i^2 = -28i + 7(-1) = \boxed{-7 - 28i}$

(7) $(1-i)(-1+i) = 1(-1) + 1(i) - i(-1) - i(i) = -1 + i + i - i^2 =$
$-1 + i + i - (-1) = -1 + 2i + 1 = \boxed{2i}$ Note: This answer is purely imaginary.

(8) $(4+6i)(4+6i) = 4(4) + 4(6i) + 6i(4) + 6i(6i) = 16 + 24i + 24i + 36(-1) =$
$16 + 48i - 36 = \boxed{-20 + 48i}$

(9) $(7+9i)(7-9i) = 7(7) + 7(-9i) + 9i(7) + 9i(-9i) = 49 - 63i + 63i - 81(-1) =$
$49 + 81 = \boxed{130}$ Note: This answer is purely real. See Sec. 3.5.

(10) $(8-5i)^2 = (8-5i)(8-5i) = 8(8) + 8(-5i) - 5i(8) - 5i(-5i) =$
$64 - 40i - 40i + 25(-1) = 64 - 80i - 25 = \boxed{39 - 80i}$

(11) $(3+2i)^3 = (3+2i)(3+2i)(3+2i) = (3+2i)[3(3) + 3(2i) + 2i(3) + 2i(2i)] =$
$(3+2i)[9 + 6i + 6i + 4(-1)] = (3+2i)(9 + 12i - 4) = (3+2i)(5+12i) =$
$= 3(5) + 3(12i) + 2i(5) + 2i(12i) = 15 + 36i + 10i + 24(-1) = 15 + 46i - 24 =$
$\boxed{-9 + 46i}$

(12) $i^{46} = i^{44}i^2 = (i^4)^{11}i^2 = (1)^{11}(-1) = 1(-1) = \boxed{-1}$
Alternative solution: Since $46 \div 4$ has a remainder of 2, $i^{46} = i^2 = -1$.

(13) $i^{101} = i^{100}i^1 = (i^4)^{25}i^1 = (1)^{25}(i) = 1(i) = \boxed{i}$
Alternative solution: Since $101 \div 4$ has a remainder of 1, $i^{101} = i^1 = i$.

(14) $i^{75} = i^{72}i^3 = (i^4)^{18}i^3 = (1)^{18}(-i) = 1(-i) = \boxed{-i}$
Alternative solution: Since $75 \div 4$ has a remainder of 3, $i^{75} = i^3 = -i$.

(15) $i^{68} = (i^4)^{17} = (1)^{17} = \boxed{1}$
Alternative solution: Since $68 \div 4$ has a remainder of 0, $i^{68} = i^0 = i^4 = 1$.

3.5 Problems

(1) $z^* = 9 - 7i$

(2) $w^* = 5 + 8i$

(3) $u^* = -1 - i$

(4) $z^* = -3 + 6i$

(5) $v^* = -16i$

(6) $t^* = 4 + i$

(7) $z^* = \sqrt{2} - i\sqrt{3}$

(8) $w^* = 3.25 - 1.75i$

(9) $p^* = 3 - \sqrt{5} - 6i + i\sqrt{5}$

(10) $q^* = i$

(11) $d^* = 12$

(12) $z^* = \frac{1}{2} - \frac{i\sqrt{3}}{2}$

3.6 Problems

(1) $|z| = \sqrt{(8 + 6i)(8 - 6i)} = \sqrt{8^2 + 6^2} = \sqrt{64 + 36} = \sqrt{100} = \boxed{10}$

Note: $(x + iy)(x - iy) = x^2 + x(-iy) + iy(x) + iy(-iy) = x^2 - i^2y^2 = x^2 + y^2$ because $x^2 - i^2y^2 = x^2 - (-1)y^2 = x^2 + y^2$. For every problem, simply square the real and imaginary parts and **add** these squares together, then take the square root.

(2) $|w| = \sqrt{(\sqrt{3} - i)(\sqrt{3} + i)} = \sqrt{(\sqrt{3})^2 + 1^2} = \sqrt{3 + 1} = \sqrt{4} = \boxed{2}$

Note: $(\sqrt{3} - i)(\sqrt{3} + i) = \sqrt{3}\sqrt{3} + i\sqrt{3} - i\sqrt{3} - i^2 = 3 - (-1) = 3 + 1 = 4$. As with every problem in this section, the two terms get added: $(\sqrt{3} - i)(\sqrt{3} + i) = (\sqrt{3})^2 + 1^2$.

(3) $|u| = \sqrt{(-3 + 6i)(-3 - 6i)} = \sqrt{(-3)^2 + 6^2} = \sqrt{9 + 36} = \sqrt{45} = \sqrt{9(5)} = \sqrt{9}\sqrt{5} = \boxed{3\sqrt{5}}$

(4) $|z| = \sqrt{(-\sqrt{2} - i\sqrt{2})(-\sqrt{2} + i\sqrt{2})} = \sqrt{(-\sqrt{2})^2 + (\sqrt{2})^2} = \sqrt{2 + 2} = \sqrt{4} = \boxed{2}$

Note: Whether you write $\sqrt{(-\sqrt{2})^2 + (\sqrt{2})^2}$, $\sqrt{(-\sqrt{2})^2 + (-\sqrt{2})^2}$, or $\sqrt{(\sqrt{2})^2 + (\sqrt{2})^2}$ makes no difference in this or the other problems because the minus sign gets squared.

(5) $|v| = \sqrt{(12 + 5i)(12 - 5i)} = \sqrt{12^2 + 5^2} = \sqrt{144 + 25} = \sqrt{169} = \boxed{13}$

(6) $|t| = \sqrt{(7i)(-7i)} = \sqrt{-49i^2} = \sqrt{-49(-1)} = \sqrt{49} = \boxed{7}$

(7) $|z| = \sqrt{(\sqrt{6} - i\sqrt{3})(\sqrt{6} + i\sqrt{3})} = \sqrt{(\sqrt{6})^2 + (\sqrt{3})^2} = \sqrt{6 + 3} = \sqrt{9} = \boxed{3}$

(8) $|w| = \sqrt{\left(-\frac{1}{2} + \frac{i}{4}\right)\left(-\frac{1}{2} - \frac{i}{4}\right)} = \sqrt{\left(-\frac{1}{2}\right)^2 + \left(\frac{1}{4}\right)^2} = \sqrt{\frac{1}{4} + \frac{1}{16}} = \sqrt{\frac{4}{16} + \frac{1}{16}} = \sqrt{\frac{5}{16}} = \boxed{\frac{\sqrt{5}}{4}}$

(9) $|p| = \sqrt{(-9-3i)(-9+3i)} = \sqrt{(-9)^2 + 3^2} = \sqrt{81+9} = \sqrt{90} = \sqrt{9}\sqrt{10} = \boxed{3\sqrt{10}}$

(10) $|q| = \sqrt{(8-15i)(8+15i)} = \sqrt{8^2 + 15^2} = \sqrt{64+225} = \sqrt{289} = \boxed{17}$

(11) $|d| = \sqrt{(\sqrt{11}+i)(\sqrt{11}-i)} = \sqrt{(\sqrt{11})^2 + 1^2} = \sqrt{11+1} = \sqrt{12} = \sqrt{4}\sqrt{3} = \boxed{2\sqrt{3}}$

(12) $|z| = \sqrt{(-5+10i)(-5-10i)} = \sqrt{(-5)^2 + 10^2} = \sqrt{25+100} = \sqrt{125} = \sqrt{25}\sqrt{5} = \boxed{5\sqrt{5}}$

(13) $|w| = \sqrt{(12-i\sqrt{6})(12+i\sqrt{6})} = \sqrt{12^2 + (\sqrt{6})^2} = \sqrt{144+6} = \sqrt{150} = \sqrt{25}\sqrt{6} = \boxed{5\sqrt{6}}$

(14) $|c| = \sqrt{(1-i)(1+i)} = \sqrt{1^2 + 1^2} = \sqrt{1+1} = \boxed{\sqrt{2}}$

Note: $(1-i)(1+i) = 1+i-i-i^2 = 1-(-1) = 1+1 = 2$. As with every problem in this section, the two terms get added: $(1-i)(1+i) = 1^2 + 1^2$.

(15) $|z| = \sqrt{(-3\sqrt{2}-4i\sqrt{5})(-3\sqrt{2}+4i\sqrt{5})} = \sqrt{(-3\sqrt{2})^2 + (4\sqrt{5})^2} =$

$\sqrt{3^2(\sqrt{2})^2 + 4^2(\sqrt{5})^2} = \sqrt{9(2)+16(5)} = \sqrt{18+80} = \sqrt{98} = \sqrt{49}\sqrt{2} = \boxed{7\sqrt{2}}$

3.7 Problems

(1) $\dfrac{2+4i}{5+3i} = \dfrac{2+4i}{5+3i}\left(\dfrac{5-3i}{5-3i}\right) = \dfrac{2(5)+2(-3i)+4i(5)+4i(-3i)}{5^2+3^2} = \dfrac{10-6i+20i-12(-1)}{25+9} = \dfrac{10+14i+12}{34} =$

$\dfrac{22+14i}{34} = \boxed{\dfrac{11+7i}{17}}$ Note: Divide $22+14i$ and 34 each by 2 to reduce $\dfrac{22+14i}{34}$ to $\dfrac{11+7i}{17}$.

(2) $\dfrac{7+3i}{2-6i} = \dfrac{7+3i}{2-6i}\left(\dfrac{2+6i}{2+6i}\right) = \dfrac{7(2)+7(6i)+3i(2)+3i(6i)}{2^2+6^2} = \dfrac{14+42i+6i+18(-1)}{4+36} = \dfrac{14+48i-18}{40} =$

$\dfrac{-4+48i}{40} = \boxed{\dfrac{-1+12i}{10}}$ Note: Divide $-4+48i$ and 40 each by 4 to reduce $\dfrac{-4+48i}{40}$ to $\dfrac{-1+12i}{10}$.

(3) $\dfrac{9-4i}{5+8i} = \dfrac{9-4i}{5+8i}\left(\dfrac{5-8i}{5-8i}\right) = \dfrac{9(5)+9(-8i)-4i(5)-4i(-8i)}{5^2+8^2} = \dfrac{45-72i-20i+32(-1)}{25+64} = \dfrac{45-92i-32}{89} = \boxed{\dfrac{13-92i}{89}}$

(4) $\dfrac{10-15i}{5-2i} = \dfrac{10-15i}{5-2i}\left(\dfrac{5+2i}{5+2i}\right) = \dfrac{10(5)+10(2i)-15i(5)-15i(2i)}{5^2+2^2} = \dfrac{50+20i-75i-30(-1)}{29} = \dfrac{50-55i+30}{29} =$

$\boxed{\dfrac{80-55i}{29}}$ Alternative answers: $\dfrac{5(16-11i)}{29} = \dfrac{5}{29}(16-11i)$.

(5) $\dfrac{1+i}{1-i} = \dfrac{1+i}{1-i}\left(\dfrac{1+i}{1+i}\right) = \dfrac{1(1)+1(i)+i(1)+i(i)}{1^2+1^2} = \dfrac{1+i+i-1}{2} = \dfrac{2i}{2} = \boxed{i}$

Note: $(1-i)(1+i) = 1+i-i-i^2 = 1-(-1) = 1+1 = 2$. As with the other problems in this section, the squares get added in the denominator: $1^2 + 1^2 = 2$.

(6) $\dfrac{6}{7+9i} = \dfrac{6}{7+9i}\left(\dfrac{7-9i}{7-9i}\right) = \dfrac{6(7)+6(-9i)}{7^2+9^2} = \dfrac{42-54i}{49+81} = \dfrac{42-54i}{130} = \boxed{\dfrac{21-27i}{65}}$ (You could factor out a 3...)

(7) $\dfrac{6+3i}{4-8i} = \dfrac{6+3i}{4-8i}\left(\dfrac{4+8i}{4+8i}\right) = \dfrac{6(4)+6(8i)+3i(4)+3i(8i)}{4^2+8^2} = \dfrac{24+48i+12i+24(-1)}{16+64} = \dfrac{24+60i-24}{80} = \dfrac{60i}{80} = \boxed{\dfrac{3i}{4}}$

Note: Divide $60i$ and 80 each by 20 to reduce $\dfrac{60i}{80}$ to $\dfrac{3i}{4}$.

(8) $\dfrac{2-i}{3-7i} = \dfrac{2-i}{3-7i}\left(\dfrac{3+7i}{3+7i}\right) = \dfrac{2(3)+2(7i)-i(3)-i(7i)}{3^2+7^2} = \dfrac{6+14i-3i-7(-1)}{9+49} = \dfrac{6+11i+7}{58} = \boxed{\dfrac{13+11i}{58}}$

(9) $\dfrac{8}{9-3i} = \dfrac{8}{9-3i}\left(\dfrac{9+3i}{9+3i}\right) = \dfrac{8(9)+8(3i)}{9^2+3^2} = \dfrac{72+24i}{81+9} = \dfrac{72+24i}{90} = \boxed{\dfrac{12+4i}{15}}$

Note: Divide $72+24i$ and 90 each by 6 to reduce $\dfrac{72+24i}{90}$ to $\dfrac{12+4i}{15}$.

(10) $\dfrac{6-5i}{8+9i} = \dfrac{6-5i}{8+9i}\left(\dfrac{8-9i}{8-9i}\right) = \dfrac{6(8)+6(-9i)-5i(8)-5i(-9i)}{8^2+9^2} = \dfrac{48-54i-40i+45(-1)}{64+81} = \dfrac{48-94i-45}{145} = \boxed{\dfrac{3-94i}{145}}$

3.8 Problems

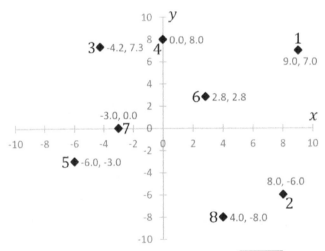

(1) $z = \sqrt{(9+7i)(9-7i)} = \sqrt{9^2+7^2} = \sqrt{81+49} = \boxed{\sqrt{130}}$

(2) $w = \sqrt{(8-6i)(8+6i)} = \sqrt{8^2+6^2} = \sqrt{64+36} = \sqrt{100} = \boxed{10}$

Note: Whether you write $\sqrt{8^2+(-6)^2}$ or $\sqrt{8^2+6^2}$ makes no difference since $(-1)^2 = 1$.

(3) $u = \sqrt{\left(-3\sqrt{2}+3i\sqrt{6}\right)\left(-3\sqrt{2}-3i\sqrt{6}\right)} = \sqrt{\left(-3\sqrt{2}\right)^2+\left(3\sqrt{6}\right)^2} = \sqrt{3^2(2)+3^2(6)} =$

$\sqrt{9(2)+9(6)} = \sqrt{18+54} = \sqrt{72} = \sqrt{36}\sqrt{2} = \boxed{6\sqrt{2}}$

(4) $v = \sqrt{(0+8i)(0-8i)} = \sqrt{0^2+8^2} = \sqrt{0+64} = \sqrt{64} = \boxed{8}$

(5) $t = \sqrt{(-6-3i)(-6+3i)} = \sqrt{(-6)^2 + 3^2} = \sqrt{36+9} = \sqrt{45} = \sqrt{9}\sqrt{5} = \boxed{3\sqrt{5}}$

(6) $c = \sqrt{(2\sqrt{2}+2i\sqrt{2})(2\sqrt{2}-2i\sqrt{2})} = \sqrt{(2\sqrt{2})^2 + (2\sqrt{2})^2} = \sqrt{2^2(2) + 2^2(2)} =$

$\sqrt{4(2)+4(2)} = \sqrt{8+8} = \sqrt{16} = \boxed{4}$

(7) $d = \sqrt{(-3+0i)(-3-0i)} = \sqrt{(-3)^2 + 0^2} = \sqrt{9} = \boxed{3}$

(8) $q = \sqrt{(4-8i)(4+8i)} = \sqrt{4^2 + 8^2} = \sqrt{16+64} = \sqrt{80} = \sqrt{16}\sqrt{5} = \boxed{4\sqrt{5}}$

3.9 Problems

(1) $f(6) = \sqrt{11-6^2} = \sqrt{11-36} = \sqrt{-25} = \boxed{5i}$

(2) $g(7-5i) = 4(7-5i) + 9 = 4(7) + 4(-5i) + 9 = 28 - 20i + 9 = \boxed{37-20i}$

(3) $y(2i) = 8(2i)^2 - 5i(2i) = 8(2)^2 i^2 - 10i^2 = 8(4)(-1) - 10(-1) = -32 + 10 = \boxed{-22}$

(4) $p(9-6i) = 2(9-6i)^2 = 2(9-6i)(9-6i) = 2[9(9) + 9(-6i) - 6i(9) - 6i(-6i)] =$

$2[81 - 54i - 54i + 36(-1)] = 2(81 - 108i - 36) = 2(45 - 108i) = \boxed{90-216i}$

Alternative answer: $18(5-12i)$.

(5) $h(5+2i) = \frac{2(5+2i)-5i}{4(5+2i)+3i} = \frac{10+4i-5i}{20+8i+3i} = \frac{10-i}{20+11i} = \frac{10-i}{20+11i}\left(\frac{20-11i}{20-11i}\right) = \frac{10(20)+10(-11i)-i(20)-i(-11i)}{20^2+11^2}$

$= \frac{200-110i-20i+11(-1)}{400+121} = \frac{200-130i-11}{521} = \boxed{\frac{189-130i}{521}}$

Note: We multiplied the denominator by the complex conjugate (Sec. 3.7) in order to express the answer in standard form.

(6) $f(-16) = (-16)^{3/2} = \left[(-16)^{1/2}\right]^3 = \left(\sqrt{-16}\right)^3 = (4i)^3 = 4^3 i^3 = 64(-i) = \boxed{-64i}$

Recall Sec. 2.6 and Sec. 2.11.

3.10 Problems

(1) $x = \boxed{\pm 6i}$ Check: $76 - 5(\pm 6i)^2 = 76 - 5(6)^2 i^2 = 76 - 5(36)(-1) = 76 + 180 = 256$.

(2) $z = \boxed{-7-9i}$ Check: $9(-7-9i) + 14 = -63 - 81i + 14 = -49 - 81i$ agrees

with $7(-7-9i) - 18i = -49 - 63i - 18i = -49 - 81i$.

(3) $z = \boxed{8+6i}$ Check: $11i - 19 - 2(8+6i) = 11i - 19 - 16 - 12i = -35 - i$ agrees

with $13 + 35i - 6(8+6i) = 13 + 35i - 48 - 36i = -35 - i$.

(4) $x = \boxed{\pm 4i\sqrt{3}}$ Check: $24 - \left(\pm 4i\sqrt{3}\right)^2 = 24 - 4^2 i^2 \left(\sqrt{3}\right)^2 = 24 - 16(-1)(3) = 24 + 48 = 72$ agrees with $\left(\pm 4i\sqrt{3}\right)^2 + 120 = 4^2 i^2 \left(\sqrt{3}\right)^2 + 120 = 16(-1)(3) + 120 = -48 + 120 = 72$. Note: The answer is equivalent to $\pm i\sqrt{48}$, but $\pm i\sqrt{48}$ isn't in standard form.

(5) $x = \boxed{\pm 7i\sqrt{2}}$ Check: $350 - 4\left(\pm 7i\sqrt{2}\right)^2 = 350 - 4(7)^2 i^2 \left(\sqrt{2}\right)^2 = 350 - 4(49)(-1)(2) = 350 + 392 = 742$ agrees with $56 - 7\left(\pm 7i\sqrt{2}\right)^2 = 56 - 7(49)(-1)(2) = 56 + 686 = 742$. Note: The answer is equivalent to $\pm i\sqrt{98}$, but $\pm i\sqrt{98}$ isn't in standard form.

(6) $z = \boxed{\dfrac{13+9i}{5}}$ Check: $\dfrac{2+i}{z} = (2+i) \div \dfrac{13+9i}{5} = (2+i)\dfrac{5}{13+9i} = \dfrac{10+5i}{13+9i} = \dfrac{10+5i}{13+9i}\left(\dfrac{13-9i}{13-9i}\right) = $

$\dfrac{10(13)+10(-9i)+5i(13)+5i(-9i)}{13^2+9^2} = \dfrac{130-90i+65i-45(-1)}{169+81} = \dfrac{130-25i+45}{250} = \dfrac{175-25i}{250} = \dfrac{25(7-i)}{25(10)} = \dfrac{7-i}{10}$

agrees with $\dfrac{2-i}{3-i} = \dfrac{2-i}{3-i}\left(\dfrac{3+i}{3+i}\right) = \dfrac{2(3)+2i-3i-i^2}{3^2+1^2} = \dfrac{6-i-(-1)}{9+1} = \dfrac{6-i+1}{10} = \dfrac{7-i}{10}$.

(7) $z = \boxed{-3+6i}$ Check: $12 - 3[6i - (-3+6i)] = 12 - 3(6i+3-6i) = 12 - 3(3) = 12 - 9 = 3$ agrees with $6i - (-3+6i) = 6i + 3 - 6i = 3$.

(8) $x = \boxed{\pm 9i}$ Check: $3(\pm 9i)^2 + 300 = 3(9)^2 i^2 + 300 = 3(81)(-1) + 300 = -243 + 300 = 57$.

(9) $z = \boxed{2-3i}$ such that $z^* = 2 + 3i$. Note that z^* is the complex conjugate (Sec. 3.5) of z. Check: $6(2-3i)^* - 12i = 6(2+3i) - 12i = 12 + 18i - 12i = 12 + 6i$ agrees with $2(2-3i)^* + 8 = 2(2+3i) + 8 = 4 + 6i + 8 = 12 + 6i$.

(10) $x = \boxed{\pm 6i\sqrt{2}}$ Check: $\dfrac{18}{\pm 6i\sqrt{2}} = \dfrac{3}{\pm i\sqrt{2}}\left(\dfrac{\pm i\sqrt{2}}{\pm i\sqrt{2}}\right) = \dfrac{\pm 3i\sqrt{2}}{i^2(2)} = \dfrac{\pm 3i\sqrt{2}}{-2} = \dfrac{\mp 3i\sqrt{2}}{2}$ agrees with

$-\dfrac{\pm 6i\sqrt{2}}{4} = \dfrac{\mp 3i\sqrt{2}}{2}$ (where $\mp$ means the opposite sign compared to $\pm$).

(11) $x = \boxed{-16i}$, $x = \boxed{0}$, or $x = \boxed{16i}$ Check: $\dfrac{(\pm 16i)^3}{8} = \dfrac{\pm 16^3 i^3}{8} = \dfrac{\pm 4096(-i)}{8} = \mp 512i$ agrees with $-32(\pm 16i) = \mp 512i$. To get $x = 0$ as a solution, factor the given equation to get $\dfrac{x^3}{8} + 32x = x\left(\dfrac{x^2}{8} + 32\right) = 0$, such that either $x = 0$ or $\dfrac{x^2}{8} + 32 = 0$.

(12) $x = \boxed{\pm 10i}$ Check: $2(\pm 10i)^2 + 350 = 2(10)^2 i^2 + 350 = 2(100)(-1) + 350 = 150$.

(13) $z = \boxed{\dfrac{-12+18i}{13}} = \boxed{\dfrac{6}{13}(-2+3i)}$ Check: $\dfrac{1}{2i} - \dfrac{13}{6(-2+3i)} = \dfrac{3(-2+3i)}{6i(-2+3i)} - \dfrac{13i}{6i(-2+3i)} = $

$\dfrac{-6+9i-13i}{-12i+18(-1)} = \dfrac{-6-4i}{-18-12i} = \dfrac{-6-4i}{3(-6-4i)} = \dfrac{1}{3}$. Note: $\dfrac{1}{z} = \dfrac{13}{-12+18i}$ is equivalent to $z = \dfrac{-12+18i}{13}$.

(14) $z = \boxed{-9i}$ Check: $\dfrac{3}{-9i+3i} = \dfrac{3}{-6i} = -\dfrac{1}{2i}$ agrees with $\dfrac{7}{-9i-5i} = \dfrac{7}{-14i} = -\dfrac{1}{2i}$.

4 Quadratics

4.1 Problems

(1) $x = \boxed{5}$ or $x = \boxed{8}$. Check: $5^2 - 13(5) + 40 = 25 - 65 + 40 = 0$ and
$8^2 - 13(8) + 40 = 64 - 104 + 40 = 0$.

(2) $x = \boxed{-9}$ or $x = \boxed{\frac{2}{3}}$. Check: $3(-9)^2 + 25(-9) = 3(81) - 225 = 243 - 225 = 18$
and $3\left(\frac{2}{3}\right)^2 + 25\left(\frac{2}{3}\right) = 3\left(\frac{4}{9}\right) + \frac{50}{3} = \frac{4}{3} + \frac{50}{3} = \frac{54}{3} = 18$.

(3) $x = \boxed{1 - \sqrt{3}} \approx \boxed{-0.7321}$ or $x = \boxed{1 + \sqrt{3}} \approx \boxed{2.732}$. Calculator check:
$2(-0.7321) - 6 + 3(-0.7321)^2 \approx -5.856$ agrees with $8(-0.7321) \approx -5.857$
(within reasonable round-off error; to reduce round-off error, keep more digits) and
$2(2.732) - 6 + 3(2.732)^2 \approx 21.86$ agrees with $8(2.732) \approx 21.86$.

(4) $x = \boxed{-4}$ or $x = \boxed{-\frac{7}{4}}$. Check: $30(-4) + 4(-4)^2 = -120 + 4(16) = -120 + 64 =$
-56 agrees with $7(-4) - 28 = -28 - 28 = -56$ and $30\left(-\frac{7}{4}\right) + 4\left(-\frac{7}{4}\right)^2 =$
$-\frac{210}{4} + 4\left(\frac{49}{16}\right) = -\frac{210}{4} + \frac{49}{4} = \frac{-161}{4}$ agrees with $7\left(-\frac{7}{4}\right) - 28 = -\frac{49}{4} - \frac{112}{4} = -\frac{161}{4}$.

(5) $x = \boxed{\frac{1}{6}}$ or $x = \boxed{\frac{5}{3}}$. Check: $33\left(\frac{1}{6}\right) - 9 = \frac{33}{6} - \frac{54}{6} = -\frac{21}{6} = -\frac{7}{2}$ agrees with $18\left(\frac{1}{6}\right)^2 - 4 =$
$18\left(\frac{1}{36}\right) - 4 = \frac{1}{2} - \frac{8}{2} = -\frac{7}{2}$ and $33\left(\frac{5}{3}\right) - 9 = 55 - 9 = 46$ agrees with $18\left(\frac{5}{3}\right)^2 - 4 =$
$18\left(\frac{25}{9}\right) - 4 = 50 - 4 = 46$.

(6) $x = \boxed{-3 - \sqrt{5}} \approx \boxed{-5.236}$ or $x = \boxed{-3 + \sqrt{5}} \approx \boxed{-0.7639}$. Calculator check:
$12(-5.236) + 2(-5.236)^2 \approx -8$ and $12(-0.7639) + 2(-0.7639)^2 \approx -8$.

(7) $x = \boxed{-2 - \sqrt{2}} \approx \boxed{-3.414}$ or $x = \boxed{-2 + \sqrt{2}} \approx \boxed{-0.5858}$. Calculator check:
$5(-3.414)^2 + 24(-3.414) \approx -23.66$ agrees with $4(-3.414) - 10 \approx -23.66$ and
$5(-0.5858)^2 + 24(-0.5858) \approx -12.34$ agrees with $4(-0.5858) - 10 \approx -12.34$.

(8) $x = \boxed{-\frac{3}{4}}$ or $x = \boxed{12}$. Check: $9\left(-\frac{3}{4}\right)^2 - 45\left(-\frac{3}{4}\right) = 9\left(\frac{9}{16}\right) + \frac{135}{4} = \frac{81}{16} + \frac{540}{16} = \frac{621}{16}$
agrees with $5\left(-\frac{3}{4}\right)^2 + 36 = 5\left(\frac{9}{16}\right) + 36 = \frac{45}{16} + \frac{576}{16} = \frac{621}{16}$ and $9(12)^2 - 45(12) =$
$1296 - 540 = 756$ agrees with $5(12)^2 + 36 = 720 + 36 = 756$.

(9) $x = \boxed{3 - \sqrt{6}} \approx \boxed{0.5505}$ or $x = \boxed{3 + \sqrt{6}} \approx \boxed{5.449}$. Calculator check:

$4(0.5505)^2 + 3 \approx 4.212$ agrees with $3(0.5505)(0.5505 + 2) \approx 4.212$ and

$4(5.449)^2 + 3 \approx 121.8$ agrees with $3(5.449)(5.449 + 2) \approx 121.8$.

(10) $x = \boxed{-\frac{9}{8}}$ or $x = \boxed{\frac{2}{5}}$. Check: $11\left(-\frac{9}{8}\right) + 18 = -\frac{99}{8} + \frac{144}{8} = \frac{45}{8}$ agrees with

$40\left(-\frac{9}{8}\right)^2 + 40\left(-\frac{9}{8}\right) = 40\left(\frac{81}{64}\right) - \frac{360}{8} = \frac{405}{8} - \frac{360}{8} = \frac{45}{8}$ and $11\left(\frac{2}{5}\right) + 18 = \frac{22}{5} + \frac{90}{5} =$

$\frac{112}{5}$ agrees with $40\left(\frac{2}{5}\right)^2 + 40\left(\frac{2}{5}\right) = 40\left(\frac{4}{25}\right) + \frac{80}{5} = \frac{32}{5} + \frac{80}{5} = \frac{112}{5}$.

(11) $x = \boxed{-3}$ or $x = \boxed{-\frac{4}{9}}$. Check: $15(-3)^2 + 31(-3) = 15(9) - 93 = 135 - 93 = 42$

agrees with $6(-3)^2 - 12 = 6(9) - 12 = 54 - 12 = 42$ and $15\left(-\frac{4}{9}\right)^2 + 31\left(-\frac{4}{9}\right) =$

$15\left(\frac{16}{81}\right) - \frac{124}{9} = \frac{80}{27} - \frac{372}{27} = -\frac{292}{27}$ agrees with $6\left(-\frac{4}{9}\right)^2 - 12 = 6\left(\frac{16}{81}\right) - 12 = \frac{32}{27} - \frac{324}{27} = -\frac{292}{27}$.

(12) $x = \boxed{1 - \sqrt{3}} \approx \boxed{-0.7321}$ or $x = \boxed{1 + \sqrt{3}} \approx \boxed{2.732}$. Calculator check:

$[2(-0.7321) + 8](-0.7321 - 6) \approx (6.536)(-6.732) \approx -44$ and

$[2(2.732) + 8](2.732 - 6) \approx (13.46)(-3.268) \approx -44$.

4.2 Problems

(1) $x^2 + 8x - 9 = \boxed{(x + 9)(x - 1)}$

Check: $(x + 9)(x - 1) = x^2 - x + 9x - 9 = x^2 + 8x - 9$.

(2) $x^2 + 13x + 30 = \boxed{(x + 10)(x + 3)}$

Check: $(x + 10)(x + 3) = x^2 + 3x + 10x + 30 = x^2 + 13x + 30$.

(3) $x^2 - 8x + 16 = \boxed{(x - 4)(x - 4)}$

Check: $(x - 4)(x - 4) = x^2 - 4x - 4x + 16 = x^2 - 8x + 16$.

(4) $x^2 - 7x - 18 = \boxed{(x - 9)(x + 2)}$

Check: $(x - 9)(x + 2) = x^2 + 2x - 9x - 18 = x^2 - 7x - 18$.

(5) $x^2 + 9x - 36 = \boxed{(x + 12)(x - 3)}$

Check: $(x + 12)(x - 3) = x^2 - 3x + 12x - 36 = x^2 + 9x - 36$.

(6) $x^2 + 12x + 32 = \boxed{(x + 8)(x + 4)}$

Check: $(x + 8)(x + 4) = x^2 + 4x + 8x + 32 = x^2 + 12x + 32$.

(7) $x^2 - 36 = \boxed{(x+6)(x-6)}$

Check: $(x+6)(x-6) = x^2 - 6x + 6x - 36 = x^2 - 36$.

(8) $x^2 - 14x + 40 = \boxed{(x-10)(x-4)}$

Check: $(x-10)(x-4) = x^2 - 4x - 10x + 40 = x^2 - 14x + 40$.

(9) $x^2 - 7x - 60 = \boxed{(x-12)(x+5)}$

Check: $(x-12)(x+5) = x^2 + 5x - 12x - 60 = x^2 - 7x - 60$.

(10) $x^2 + 17x + 16 = \boxed{(x+16)(x+1)}$

Check: $(x+16)(x+1) = x^2 + x + 16x + 16 = x^2 + 17x + 16$.

(11) $x^2 - 16x + 28 = \boxed{(x-14)(x-2)}$

Check: $(x-14)(x-2) = x^2 - 2x - 14x + 28 = x^2 - 16x + 28$.

(12) $x^2 + 17x - 60 = \boxed{(x+20)(x-3)}$

Check: $(x+20)(x-3) = x^2 - 3x + 20x - 60 = x^2 + 17x - 60$.

4.3 Problems

(1) $x = \boxed{2}$ or $x = \boxed{4}$. Check: $6(2)^2 - 36(2) + 48 = 6(4) - 72 + 48 = 24 - 24 = 0$ and $6(4)^2 - 36(4) + 48 = 6(16) - 144 + 48 = 96 - 96 = 0$.

(2) $x = \boxed{-\dfrac{6}{5}}$ or $x = \boxed{3}$. Check: $15\left(-\dfrac{6}{5}\right)^2 - 27\left(-\dfrac{6}{5}\right) - 54 = 15\left(\dfrac{36}{25}\right) + \dfrac{162}{5} - 54 =$

$\dfrac{108}{5} + \dfrac{162}{5} - \dfrac{270}{5} = 0$ and $15(3)^2 - 27(3) - 54 = 15(9) - 81 - 54 = 135 - 81 - 54 = 0$.

(3) $x = \boxed{-12}$ or $x = \boxed{\dfrac{3}{8}}$. Check: $8(-12)^2 + 93(-12) - 36 = 8(144) - 1116 - 36 =$

$1152 - 1152 = 0$ and $8\left(\dfrac{3}{8}\right)^2 + 93\left(\dfrac{3}{8}\right) - 36 = 8\left(\dfrac{9}{64}\right) + \dfrac{279}{8} - 36 = \dfrac{9}{8} + \dfrac{279}{8} - \dfrac{288}{8} = 0$.

(4) $x = \boxed{-\dfrac{8}{3}}$ or $x = \boxed{-1}$. Check: $9\left(-\dfrac{8}{3}\right)^2 + 33\left(-\dfrac{8}{3}\right) + 24 = 9\left(\dfrac{64}{9}\right) - 88 + 24 =$

$64 - 64 = 0$ and $9(-1)^2 + 33(-1) + 24 = 9(1) - 33 + 24 = 9 - 9 = 0$.

(5) $x = \boxed{-\dfrac{7}{6}}$ or $x = \boxed{\dfrac{5}{2}}$. Check: $12\left(-\dfrac{7}{6}\right)^2 - 16\left(-\dfrac{7}{6}\right) - 35 = 12\left(\dfrac{49}{36}\right) + \dfrac{112}{6} - 35 =$

$\dfrac{49}{3} + \dfrac{56}{3} - \dfrac{105}{3} = 0$ and $12\left(\dfrac{5}{2}\right)^2 - 16\left(\dfrac{5}{2}\right) - 35 = 12\left(\dfrac{25}{4}\right) - 40 - 35 = 75 - 75 = 0$.

(6) $x = \boxed{\frac{5}{9}}$ or $x = \boxed{5}$. Check: $9\left(\frac{5}{9}\right)^2 - 50\left(\frac{5}{9}\right) + 25 = 9\left(\frac{25}{81}\right) - \frac{250}{9} + 25 = \frac{25}{9} - \frac{250}{9} + \frac{225}{9} =$

0 and $9(5)^2 - 50(5) + 25 = 9(25) - 250 + 25 = 225 - 225 = 0.$

(7) $x = \boxed{-\frac{9}{4}}$ or $x = \boxed{-\frac{1}{12}}$. Check: $48\left(-\frac{9}{4}\right)^2 + 112\left(-\frac{9}{4}\right) + 9 = 48\left(\frac{81}{16}\right) - 252 + 9 =$

$243 - 243 = 0$ and $48\left(-\frac{1}{12}\right)^2 + 112\left(-\frac{1}{12}\right) + 9 = 48\left(\frac{1}{144}\right) - \frac{28}{3} + 9 = \frac{1}{3} - \frac{28}{3} + \frac{27}{3} = 0.$

(8) $x = \boxed{-\frac{9}{10}}$ or $x = \boxed{\frac{5}{6}}$. Check: $60\left(-\frac{9}{10}\right)^2 + 4\left(-\frac{9}{10}\right) - 45 = 60\left(\frac{81}{100}\right) - \frac{36}{10} - 45 =$

$\frac{486}{10} - \frac{36}{10} - \frac{450}{10} = 0$ and $60\left(\frac{5}{6}\right)^2 + 4\left(\frac{5}{6}\right) - 45 = 60\left(\frac{25}{36}\right) + \frac{20}{6} - 45 = \frac{250}{6} + \frac{20}{6} - \frac{270}{6} = 0.$

(9) $x = \boxed{\frac{2}{3}}$ or $x = \boxed{6}$. Check: $9\left(\frac{2}{3}\right)^2 - 60\left(\frac{2}{3}\right) + 36 = 9\left(\frac{4}{9}\right) - 40 + 36 = 4 - 4 = 0$

and $9(6)^2 - 60(6) + 36 = 9(36) - 360 + 36 = 324 - 324 = 0.$

(10) $x = \boxed{-\frac{14}{5}}$ or $x = \boxed{-\frac{3}{2}}$. Check: $10\left(-\frac{14}{5}\right)^2 + 43\left(-\frac{14}{5}\right) + 42 = 10\left(\frac{196}{25}\right) - \frac{602}{5} + 42 =$

$\frac{392}{5} - \frac{602}{5} + \frac{210}{5} = 0$ and $10\left(-\frac{3}{2}\right)^2 + 43\left(-\frac{3}{2}\right) + 42 = 10\left(\frac{9}{4}\right) - \frac{129}{2} + 42 = \frac{45}{2} - \frac{129}{2} + \frac{84}{2} = 0.$

(11) $x = \boxed{-2}$ or $x = \boxed{\frac{4}{3}}$. Check: $6(-2)^2 + 4(-2) - 16 = 6(4) - 8 - 16 = 24 - 24 = 0$

and $6\left(\frac{4}{3}\right)^2 + 4\left(\frac{4}{3}\right) - 16 = 6\left(\frac{16}{9}\right) + \frac{16}{3} - 16 = \frac{32}{3} + \frac{16}{3} - \frac{48}{3} = 0.$

(12) $x = \boxed{-\frac{24}{5}}$ or $x = \boxed{\frac{3}{2}}$. Check: $10\left(-\frac{24}{5}\right)^2 + 33\left(-\frac{24}{5}\right) - 72 = 10\left(\frac{576}{25}\right) - \frac{792}{5} - 72 =$

$\frac{1152}{5} - \frac{792}{5} - \frac{360}{5} = 0$ and $10\left(\frac{3}{2}\right)^2 + 33\left(\frac{3}{2}\right) - 72 = 10\left(\frac{9}{4}\right) + \frac{99}{2} - 72 = \frac{45}{2} + \frac{99}{2} - \frac{144}{2} = 0.$

4.4 Problems

(1) $(x + 8)^2 - 55$. Check: $(x + 8)^2 - 55 = x^2 + 16x + 64 - 55 = x^2 + 16x + 9$

(2) $(4x + 3)^2 + 11$. Check: $(4x + 3)^2 + 11 = 16x^2 + 24x + 9 + 11 = 16x^2 + 24x + 20$

(3) $(x - 6)^2 - 18$. Check: $(x - 6)^2 - 18 = x^2 - 12x + 36 - 18 = x^2 - 12x + 18$

(4) $(7x + 9)^2 + 9$. Check: $(7x + 9)^2 + 9 = 49x^2 + 126x + 81 + 9 = 49x^2 + 126x + 90$

(5) $(2x - 5)^2 - 31$. Check: $(2x - 5)^2 - 31 = 4x^2 - 20x + 25 - 31 = 4x^2 - 20x - 6$

(6) $(5x - 4)^2 + 34$. Check: $(5x - 4)^2 + 34 = 25x^2 - 40x + 16 + 34 = 25x^2 - 40x + 50$

(7) $(8x + 3)^2 - 13$. Check: $(8x + 3)^2 - 13 = 64x^2 + 48x + 9 - 13 = 64x^2 + 48x - 4$

(8) $(3x - 5)^2 + 17$. Check: $(3x - 5)^2 + 17 = 9x^2 - 30x + 25 + 17 = 9x^2 - 30x + 42$

(9) $(6x + 7)^2 - 1$. Check: $(6x + 7)^2 - 1 = 36x^2 + 84x + 49 - 1 = 36x^2 + 84x + 48$

(10) $(9x - 2)^2 - 8$. Check: $(9x - 2)^2 - 8 = 81x^2 - 36x + 4 - 8 = 81x^2 - 36x - 4$

(11) $(7x - 8)^2 + 12$. Check: $(7x - 8)^2 + 12 = 49x^2 - 112x + 64 + 12 = 49x^2 - 112x + 76$

(12) $(4x + 9)^2 - 99$. Check: $(4x + 9)^2 - 99 = 16x^2 + 72x + 81 - 99 = 16x^2 + 72x - 18$

4.5 Problems

(1) Since $b^2 - 4ac = (-7)^2 - 4(2)(6) = 49 - 48 = 1 > 0$, $2x^2 - 7x + 6 = 0$ has **two** distinct **real** answers.

(2) Since $b^2 - 4ac = 5^2 - 4(1)(7) = 25 - 28 = -3 < 0$, $x^2 + 5x + 7 = 0$ has **two** distinct **complex** answers.

(3) Since $b^2 - 4ac = 12^2 - 4(8)(-5) = 144 + 160 = 304 > 0$, $8x^2 + 12x - 5 = 0$ has **two** distinct **real** answers.

(4) Since $b^2 - 4ac = (-20)^2 - 4(4)(25) = 400 - 400 = 0$, $4x^2 - 20x + 25 = 0$ has **one** distinct **real** answer.

(5) Since $b^2 - 4ac = 10^2 - 4(-3)(-8) = 100 - 96 = 4 > 0$, $-3x^2 + 10x - 8 = 0$ has **two** distinct **real** answers. Note: $-4(-3)(-8) = -96$ since there are 3 minuses.

(6) Since $b^2 - 4ac = (-2)^2 - 4(1)(-1) = 4 + 4 = 8 > 0$, $x^2 - 2x - 1 = 0$ has **two** distinct **real** answers.

(7) Since $b^2 - 4ac = 14^2 - 4(7)(7) = 196 - 196 = 0$, $7x^2 + 14x + 7 = 0$ has **one** distinct **real** answer.

(8) Since $b^2 - 4ac = (-9)^2 - 4(2)(11) = 81 - 88 = -7 < 0$, $2x^2 - 9x + 11 = 0$ has **two** distinct **complex** answers.

(9) Since $b^2 - 4ac = (-8)^2 - 4(-8)(2) = 64 + 64 = 128 > 0$, $-8x^2 - 8x + 2 = 0$ has **two** distinct **real** answers.

(10) Since $b^2 - 4ac = (-6)^2 - 4(-5)(-2) = 36 - 40 = -4 < 0$, $-5x^2 - 6x - 2 = 0$ has **two** distinct **complex** answers. Note: $-4(-5)(-2) = -40$ since there are 3 minuses.

4.6 Problems

(1) $x = \boxed{1 \pm i}$. Check: $(1 + i)^2 - 2(1 + i) + 2 = 1 + 2i - 1 - 2 - 2i + 2 = 0$.

Note: For this section, the checks only show that one of the roots solves the equation. The check for the complex conjugate is nearly identical, except for a sign change.

(2) $x = \boxed{3 \pm 2i}$. Check: $(3 + 2i)^2 - 6(3 + 2i) + 13 = 9 + 12i + 4(-1) - 18 - 12i + 13 =$

$9 - 4 - 18 + 13 = 22 - 22 = 0$.

(3) $x = \boxed{-2 \pm 5i}$. Check: $(-2 + 5i)^2 + 4(-2 + 5i) + 29 = 4 - 20i + 25(-1) - 8 + 20i + 29$

$= 4 - 25 - 8 + 29 = 33 - 33 = 0$. Note: Recall that $(-p + q)^2 = p^2 - 2pq + q^2$.

(4) $x = \boxed{\frac{1}{2} \pm i}$. Check: $-4 \left(\frac{1}{2} + i \right)^2 + 4 \left(\frac{1}{2} + i \right) - 5 = -4 \left(\frac{1}{4} + i - 1 \right) + 2 + 4i - 5 =$

$-1 - 4i + 4 + 2 + 4i - 5 = 6 - 6 = 0$.

(5) $x = \boxed{6 \pm 2i}$. Check: $2(6 + 2i)^2 - 24(6 + 2i) + 80 =$

$2(36 + 24i - 4) - 144 - 48i + 80 = 72 + 48i - 8 - 144 - 48i + 80 = 152 - 152 = 0$.

(6) $x = \boxed{-1 \pm \frac{i}{2}}$. Check: $4 \left(-1 + \frac{i}{2} \right)^2 + 8 \left(-1 + \frac{i}{2} \right) + 5 = 4 \left(1 - i - \frac{1}{4} \right) - 8 + 4i + 5 =$

$4 - 4i - 1 - 8 + 4i + 5 = 9 - 9 = 0$. Note: $(-p + q)^2 = p^2 - 2pq + q^2$. Let $p = 1$ and

$q = \frac{i}{2}$ to get $\left(-1 + \frac{i}{2} \right)^2 = 1^2 - 2(1) \left(\frac{i}{2} \right) + \left(\frac{i}{2} \right)^2 = 1 - i + \frac{i^2}{4} = 1 - i - \frac{1}{4}$.

(7) $x = \boxed{-\frac{3}{2} \pm 4i}$. Check: $-4 \left(-\frac{3}{2} + 4i \right)^2 - 12 \left(-\frac{3}{2} + 4i \right) - 73 =$

$-4 \left(\frac{9}{4} - 12i - 16 \right) + 18 - 48i - 73 = -9 + 48i + 64 + 18 - 48i - 73 = 82 - 82 = 0$.

(8) $x = \boxed{5 \pm 3i}$. Check: $3(5 + 3i)^2 - 30(5 + 3i) + 102 =$

$3(25 + 30i - 9) - 150 - 90i + 102 = 75 + 90i - 27 - 150 - 90i + 102 = 177 - 177 = 0$.

(9) $x = \boxed{\frac{2}{3} \pm \frac{1}{3}i} = \boxed{\frac{2 \pm i}{3}}$. Check: $-9 \left(\frac{2+i}{3} \right)^2 + 12 \left(\frac{2+i}{3} \right) - 5 = -9 \left(\frac{4+4i-1}{9} \right) + 4(2 + i) - 5 =$

$-4 - 4i + 1 + 8 + 4i - 5 = 9 - 9 = 0$.

(10) $x = \boxed{-2 \pm \frac{2i}{5}}$. Check: $25 \left(-2 + \frac{2i}{5} \right)^2 + 100 \left(-2 + \frac{2i}{5} \right) + 104 =$

$25 \left(4 - \frac{8i}{5} - \frac{4}{25} \right) - 200 + 40i + 104 = 100 - 40i - 4 - 200 + 40i + 104 = 204 - 204 = 0$.

(11) $x = \boxed{7 \pm 3i}$. Check: $2(7 + 3i)^2 - 28(7 + 3i) + 116 =$

$2(49 + 42i - 9) - 196 - 84i + 116 = 98 + 84i - 18 - 196 - 84i + 116 = 214 - 214 = 0$.

(12) $x = \boxed{-\frac{5}{2} \pm \frac{9}{2}i} = \boxed{\frac{-5 \pm 9i}{2}}$. Check: $-6\left(\frac{-5+9i}{2}\right)^2 - 30\left(\frac{-5+9i}{2}\right) - 159 =$

$-6\left(\frac{25-90i-81}{4}\right) - 15(-5 + 9i) - 159 = -\frac{3}{2}(-56 - 90i) + 75 - 135i - 159 =$

$84 + 135i + 75 - 135i - 159 = 159 - 159 = 0.$

Tip: Divide both sides of $-6x^2 - 30x - 159 = 0$ by 3 to make the numbers simpler.

4.7 Problems

(1) $x = \boxed{-4}$ or $x = \boxed{8}$. Check: $\frac{-4}{4} = -1$ agrees with $\frac{-4+8}{-4} = \frac{4}{-4} = -1$ and $\frac{8}{4} = 2$ agrees

with $\frac{8+8}{8} = \frac{16}{8} = 2.$

(2) $x = \boxed{\frac{1}{3}}$ or $x = \boxed{6}$. Check: $\frac{(1/3)^2+2}{1/3} = \frac{\frac{1}{9}+2}{1/3} = \frac{\frac{1}{9}+\frac{18}{9}}{1/3} = \frac{19/9}{1/3} = \frac{19}{9} \div \frac{1}{3} = \frac{19}{9} \times \frac{3}{1} = \frac{57}{9} = \frac{19}{3}$

and $\frac{6^2+2}{6} = \frac{36+2}{6} = \frac{38}{6} = \frac{19}{3}.$

(3) $x = \boxed{-4}$ or $x = \boxed{-3}$. Check: $\frac{8}{x+7} = \frac{8}{-4+7} = \frac{8}{3}$ agrees with $-\frac{2(-4)}{3} = \frac{8}{3}$ and $\frac{8}{-3+7} =$

$\frac{8}{4} = 2$ agrees with $-\frac{2(-3)}{3} = 2.$

(4) $x = \boxed{-9}$ or $x = \boxed{4}$. Check: $\frac{-9+3}{3} = \frac{-6}{3} = -2$ agrees with $\frac{14}{-9+2} = \frac{14}{-7} = -2$ and

$\frac{4+3}{3} = \frac{7}{3}$ agrees with $\frac{14}{4+2} = \frac{14}{6} = \frac{7}{3}.$

(5) $x = \boxed{-2}$ or $x = \boxed{\frac{1}{8}}$. Check: $\frac{2}{9(-2)} = \frac{2}{-18} = -\frac{1}{9}$ agrees with $\frac{5}{3-12(-2)^2} = \frac{5}{3-12(4)} = \frac{5}{3-48} =$

$\frac{5}{-45} = -\frac{1}{9}$ and $\frac{2}{9\left(\frac{1}{8}\right)} = \frac{2}{1} \div \frac{9}{8} = \frac{2}{1} \times \frac{8}{9} = \frac{16}{9}$ agrees with $\frac{5}{3-12\left(\frac{1}{8}\right)^2} = \frac{5}{3-\frac{12}{64}} = \frac{5}{\frac{192}{64}-\frac{12}{64}} = \frac{5}{\frac{180}{64}} =$

$\frac{5}{1} \times \frac{64}{180} = \frac{320}{180} = \frac{16}{9}.$

(6) $x = \boxed{\frac{2}{3}}$ or $x = \boxed{3}$. Check: $\frac{3\left(\frac{2}{3}\right)}{4\left(\frac{2}{3}\right)-3} = \frac{2}{\frac{8}{3}-\frac{9}{3}} = \frac{2}{-1/3} = -6$ agrees with $\frac{2}{\frac{2}{3}-1} = \frac{2}{-1/3} = -6$

and $\frac{3(3)}{4(3)-3} = \frac{9}{12-3} = \frac{9}{9} = 1$ agrees with $\frac{2}{3-1} = \frac{2}{2} = 1.$

(7) $x = \boxed{-\frac{5}{3}}$ or $x = \boxed{\frac{9}{2}}$. Check: $\frac{-5/3}{2(-5/3)^2-15} = \frac{-5/3}{2\left(\frac{25}{9}\right)-15} = \frac{-5/3}{\frac{50}{9}-\frac{135}{9}} = \frac{-5/3}{-\frac{85}{9}} = \frac{5}{3} \times \frac{9}{85} = \frac{3}{17}$ and

$\frac{9/2}{2(9/2)^2-15} = \frac{9/2}{2\left(\frac{81}{4}\right)-15} = \frac{9/2}{\frac{81}{2}-\frac{30}{2}} = \frac{9/2}{51/2} = \frac{9}{51} = \frac{3}{17}.$

(8) $x = \boxed{-\dfrac{3}{4}}$ or $x = \boxed{-\dfrac{1}{8}}$. Check: $\dfrac{1}{8\left(-\frac{3}{4}\right)+7} = \dfrac{1}{-6+7} = \dfrac{1}{1} = 1$ agrees with $-\dfrac{4}{3}\left(-\dfrac{3}{4}\right) = 1$

and $\dfrac{1}{8\left(-\frac{1}{8}\right)+7} = \dfrac{1}{-1+7} = \dfrac{1}{6}$ agrees with $-\dfrac{4}{3}\left(-\dfrac{1}{8}\right) = \dfrac{4}{24} = \dfrac{1}{6}$.

(9) $x = \boxed{3 - 2\sqrt{3}} \approx \boxed{-0.4641}$ or $x = \boxed{3 + 2\sqrt{3}} \approx \boxed{6.464}$. Calculator check:

$\dfrac{-0.4641-3}{6} \approx -0.5774$ agrees with $\dfrac{2}{-0.4641-3} \approx -0.5774$ and $\dfrac{6.464-3}{6} \approx 0.5774$ agrees

with $\dfrac{2}{6.464-3} \approx 0.5774$.

(10) $x = \boxed{5 - 2i}$ or $x = \boxed{5 + 2i}$. Check: $\dfrac{5-2i}{29}$ agrees with $\dfrac{1}{10-(5-2i)} = \dfrac{1}{10-5+2i} = \dfrac{1}{5+2i} =$

$\dfrac{1}{5+2i}\left(\dfrac{5-2i}{5-2i}\right) = \dfrac{5-2i}{5^2+2^2} = \dfrac{5-2i}{29}$ and $\dfrac{5+2i}{29}$ agrees with $\dfrac{1}{10-(5+2i)} = \dfrac{1}{10-5-2i} = \dfrac{1}{5-2i} =$

$\dfrac{1}{5-2i}\left(\dfrac{5+2i}{5+2i}\right) = \dfrac{5+2i}{5^2+2^2} = \dfrac{5+2i}{29}$. Notes: It may help to review Sec. 3.7. Be careful

distributing the minus sign in $10 - (5 - 2i) = 10 - 5 - (-2i) = 10 - 5 + 2i$ and in

$10 - (5 + 2i) = 10 - 5 - 2i$.

5 Systems

5.1 Problems

(1) $x = \boxed{7}$ and $y = \boxed{9}$

Check: $8(7) - 4(9) = 56 - 36 = 20$ and $2(7) + 5(9) = 14 + 45 = 59$.

(2) $x = \boxed{-4}$ and $y = \boxed{12}$

Check: $5(-4) + 3(12) = -20 + 36 = 16$ and $4(-4) + 7(12) = -16 + 84 = 68$.

(3) $x = \boxed{-5}$ and $y = \boxed{-8}$

Check: $4(-5) - 9(-8) = -20 + 72 = 52$ and $6(-5) - 3(-8) = -30 + 24 = -6$.

(4) $x = \boxed{3}$ and $y = \boxed{6}$

Check: $-5(3) + 9(6) = -15 + 54 = 39$ and $2(3) - 8(6) = 6 - 48 = -42$.

(5) $x = \boxed{\frac{7}{4}}$ and $y = \boxed{\frac{8}{3}}$

Check: $8\left(\frac{7}{4}\right) + 3\left(\frac{8}{3}\right) = 14 + 8 = 22$ and $4\left(\frac{7}{4}\right) - 6\left(\frac{8}{3}\right) = 7 - 16 = -9$.

(6) $x = \boxed{2}$ and $y = \boxed{\frac{1}{3}}$

Check: $5(2) - 6\left(\frac{1}{3}\right) = 10 - 2 = 8$ and $7(2) + 9\left(\frac{1}{3}\right) = 14 + 3 = 17$.

(7) $x = \boxed{-\frac{5}{2}}$ and $y = \boxed{\frac{3}{4}}$

Check: $7\left(-\frac{5}{2}\right) - 2\left(\frac{3}{4}\right) = -\frac{35}{2} - \frac{3}{2} = \frac{38}{2} = -19$ and $-\left(-\frac{5}{2}\right) + 6\left(\frac{3}{4}\right) = \frac{5}{2} + \frac{9}{2} = \frac{14}{2} = 7$.

(8) $x = \boxed{\frac{3}{8}}$ and $y = \boxed{\frac{5}{4}}$

Check: $8\left(\frac{3}{8}\right) + 4\left(\frac{5}{4}\right) = 3 + 5 = 8$ and $2\left(\frac{3}{8}\right) + 9\left(\frac{5}{4}\right) = \frac{3}{4} + \frac{45}{4} = \frac{48}{4} = 12$.

(9) $x = \boxed{\frac{3}{10}}$ and $y = \boxed{\frac{1}{6}}$

Check: $5\left(\frac{3}{10}\right) + 3\left(\frac{1}{6}\right) = \frac{3}{2} + \frac{1}{2} = 2$ and $10\left(\frac{3}{10}\right) - 6\left(\frac{1}{6}\right) = 3 - 1 = 2$.

(10) $x = \boxed{\frac{11}{2}}$ and $y = \boxed{-\frac{9}{2}}$

Check: $-3\left(\frac{11}{2}\right) - 7\left(-\frac{9}{2}\right) = -\frac{33}{2} + \frac{63}{2} = \frac{30}{2} = 15$ and $-7\left(\frac{11}{2}\right) + 3\left(-\frac{9}{2}\right) = -\frac{77}{2} - \frac{27}{2} = -\frac{104}{2} = -52$.

(11) $x = \boxed{-15}$ and $y = \boxed{-20}$

Check: $8(-15) - 9(-20) = -120 + 180 = 60$ and $4(-15) + 2(-20) = -60 - 40 = -100$.

(12) $x = \boxed{\frac{1}{3}}$ and $y = \boxed{\frac{5}{6}}$

Check: $9\left(\frac{1}{3}\right) - 3\left(\frac{5}{6}\right) = 3 - \frac{5}{2} = \frac{6}{2} - \frac{5}{2} = \frac{1}{2}$ and $5\left(\frac{1}{3}\right) + 2\left(\frac{5}{6}\right) = \frac{5}{3} + \frac{5}{3} = \frac{10}{3}$.

(13) $x = \boxed{9}$, $y = \boxed{5}$, and $z = \boxed{8}$

Check: $4(9) + 7(5) - 2(8) = 36 + 35 - 16 = 55$, $8(9) - 3(5) - 6(8) = 72 - 15 - 48 = 9$, and $4(5) = 20$. Note: The equation $4y = 20$ has y; it does not have x or z.

(14) $x = \boxed{6}$, $y = \boxed{2}$, and $z = \boxed{5}$

Check: $3(6) - 8(2) + 5(5) = 18 - 16 + 25 = 27$, $9(6) - 4(2) - 7(5) = 54 - 8 - 35 = 11$, and $6(2) + 3(5) = 12 + 15 = 27$. Note: $6y + 3z = 27$ does not have an x.

(15) $x = \boxed{\frac{3}{4}}$, $y = \boxed{\frac{2}{3}}$, and $z = \boxed{\frac{5}{2}}$

Check: $8\left(\frac{3}{4}\right) + 9\left(\frac{2}{3}\right) + 6\left(\frac{5}{2}\right) = 6 + 6 + 15 = 27$, $6\left(\frac{3}{4}\right) + 12\left(\frac{2}{3}\right) - 3\left(\frac{5}{2}\right) = \frac{9}{2} + 8 - \frac{15}{2} = 8 - \frac{6}{2} = 8 - 3 = 5$, and $12\left(\frac{3}{4}\right) - 6\left(\frac{2}{3}\right) - 4\left(\frac{5}{2}\right) = 9 - 4 - 10 = -5$.

(16) $x = \boxed{\frac{7}{5}}$, $y = \boxed{-\frac{3}{5}}$, and $z = \boxed{\frac{12}{5}}$

Check: $10\left(\frac{7}{5}\right) - 15\left(-\frac{3}{5}\right) + 20\left(\frac{12}{5}\right) = 14 + 9 + 48 = 71$, $25\left(\frac{7}{5}\right) + 5\left(-\frac{3}{5}\right) - 10\left(\frac{12}{5}\right) = 35 - 3 - 24 = 8$, and $15\left(\frac{7}{5}\right) - 30\left(-\frac{3}{5}\right) - 5\left(\frac{12}{5}\right) = 21 + 18 - 12 = 27$.

5.2 Problems

(1) $x = \boxed{10}$ and $y = \boxed{5}$

Check: $2(10) + 6(5) = 20 + 30 = 50$ and $7(10) - 8(5) = 70 - 40 = 30$.

(2) $x = \boxed{-6}$ and $y = \boxed{8}$

Check: $7(-6) + 9(8) = -42 + 72 = 30$ and $3(-6) + 5(8) = -18 + 40 = 22$.

(3) $x = \boxed{9}$ and $y = \boxed{4}$

Check: $7(9) - 9(4) = 63 - 36 = 27$ and $4(9) + 3(4) = 36 + 12 = 48$.

(4) $x = \boxed{8}$ and $y = \boxed{-7}$

Check: $8(8) + 7(-7) = 64 - 49 = 15$ and $-5(8) - 6(-7) = -40 + 42 = 2$.

(5) $x = \boxed{3}$ and $y = \boxed{7}$

Check: $8(3) - 12(7) = 24 - 84 = -60$ and $7(3) - 9(7) = 21 - 63 = -42$.

(6) $x = \boxed{\frac{2}{3}}$ and $y = \boxed{2}$

Check: $15\left(\frac{2}{3}\right) - 8(2) = 10 - 16 = -6$ and $12\left(\frac{2}{3}\right) + 9(2) = 8 + 18 = 26$.

(7) $x = \boxed{\frac{7}{2}}$ and $y = \boxed{-\frac{5}{4}}$

Check: $4\left(\frac{7}{2}\right) + 8\left(-\frac{5}{4}\right) = 14 - 10 = 4$ and $6\left(\frac{7}{2}\right) - 4\left(-\frac{5}{4}\right) = 21 + 5 = 26$.

(8) $x = \boxed{\frac{2}{3}}$ and $y = \boxed{\frac{3}{5}}$

Check: $-9\left(\frac{2}{3}\right) + 25\left(\frac{3}{5}\right) = -6 + 15 = 9$ and $12\left(\frac{2}{3}\right) + 35\left(\frac{3}{5}\right) = 8 + 21 = 29$.

(9) $x = \boxed{-\frac{5}{8}}$ and $y = \boxed{\frac{7}{6}}$

Check: $16\left(-\frac{5}{8}\right) + 36\left(\frac{7}{6}\right) = -10 + 42 = 32$ and $12\left(-\frac{5}{8}\right) + 21\left(\frac{7}{6}\right) = -\frac{15}{2} + \frac{49}{2} = \frac{34}{2} = 17$.

(10) $x = \boxed{\frac{1}{2}}$ and $y = \boxed{\frac{3}{2}}$

Check: $11\left(\frac{1}{2}\right) - 5\left(\frac{3}{2}\right) = \frac{11}{2} - \frac{15}{2} = -\frac{4}{2} = -2$ and $15\left(\frac{1}{2}\right) + 7\left(\frac{3}{2}\right) = \frac{15}{2} + \frac{21}{2} = \frac{36}{2} = 18$.

(11) $x = \boxed{\frac{7}{3}}$ and $y = \boxed{\frac{9}{4}}$

Check: $4\left(\frac{7}{3}\right) - 7\left(\frac{9}{4}\right) = \frac{28}{3} - \frac{63}{4} = \frac{112}{12} - \frac{189}{12} = -\frac{77}{12}$ and $8\left(\frac{7}{3}\right) - 5\left(\frac{9}{4}\right) = \frac{56}{3} - \frac{45}{4} = \frac{224}{12} - \frac{135}{12} = \frac{89}{12}$.

(12) $x = \boxed{\frac{3}{8}}$ and $y = \boxed{-\frac{5}{4}}$

Check: $-7\left(\frac{3}{8}\right) + 3\left(-\frac{5}{4}\right) = -\frac{21}{8} - \frac{15}{4} = -\frac{21}{8} - \frac{30}{8} = -\frac{51}{8}$ and $2\left(\frac{3}{8}\right) - 9\left(-\frac{5}{4}\right) = \frac{3}{4} + \frac{45}{4} = \frac{48}{4} = 12$.

5.3 Problems

(1) $\begin{vmatrix} 9 & 7 \\ 6 & 8 \end{vmatrix} = 9(8) - 7(6) = 72 - 42 = \boxed{30}$

(2) $\begin{vmatrix} 5 & 8 \\ 3 & -4 \end{vmatrix} = 5(-4) - 8(3) = -20 - 24 = \boxed{-44}$

(3) $\begin{vmatrix} 7 & 8 \\ -6 & -5 \end{vmatrix} = 7(-5) - 8(-6) = -35 + 48 = \boxed{13}$

(4) $\begin{vmatrix} 9 & 4 \\ 4 & 9 \end{vmatrix} = 9(9) - 4(4) = 81 - 16 = \boxed{65}$

(5) $\begin{vmatrix} 6 & 1 & 8 \\ 7 & 5 & 3 \\ 2 & 9 & 4 \end{vmatrix} = 6(5)(4) + 1(3)(2) + 8(7)(9) - 8(5)(2) - 1(7)(4) - 6(3)(9) =$

$120 + 6 + 504 - 80 - 28 - 162 = 630 - 270 = \boxed{360}$

(6) $\begin{vmatrix} 1 & 2 & 3 \\ 2 & 4 & 5 \\ 3 & 5 & 6 \end{vmatrix} = 1(4)(6) + 2(5)(3) + 3(2)(5) - 3(4)(3) - 2(2)(6) - 1(5)(5) =$

$24 + 30 + 30 - 36 - 24 - 25 = 84 - 85 = \boxed{-1}$

(7) $\begin{vmatrix} 2 & 3 & 4 \\ 1 & -2 & -3 \\ 0 & -1 & -5 \end{vmatrix} = 2(-2)(-5) + 3(-3)(0) + 4(1)(-1) - 4(-2)(0) - 3(1)(-5) - 2(-3)(-1)$

$= 20 - 0 - 4 + 0 + 15 - 6 = 35 - 10 = \boxed{25}$

(8) $\begin{vmatrix} 4 & 3 & 3 \\ 3 & 4 & 0 \\ 3 & 0 & 4 \end{vmatrix} = 4(4)(4) + 3(0)(3) + 3(3)(0) - 3(4)(3) - 3(3)(4) - 4(0)(0) =$

$64 + 0 + 0 - 36 - 36 - 0 = 64 - 72 = \boxed{-8}$

(9) $\begin{vmatrix} 4 & -3 & 5 \\ -2 & 1 & 4 \\ 3 & 5 & 2 \end{vmatrix} = 4(1)(2) - 3(4)(3) + 5(-2)(5) - 5(1)(3) - (-3)(-2)(2) - 4(4)(5) =$

$8 - 36 - 50 - 15 - 12 - 80 = 8 - 193 = \boxed{-185}$

(10) $\begin{vmatrix} 2 & 25 & 10 \\ 25 & 5 & 20 \\ 2 & 20 & 10 \end{vmatrix} = 2(5)(10) + 25(20)(2) + 10(25)(20) - 10(5)(2) - 25(25)(10) - 2(20)(20)$

$= 100 + 1000 + 5000 - 100 - 6250 - 800 = 6100 - 7150 = \boxed{-1050}$

(11) $\begin{vmatrix} 6 & 9 & 4 \\ 5 & 2 & 8 \\ 8 & 23 & -4 \end{vmatrix} = 6(2)(-4) + 9(8)(8) + 4(5)(23) - 4(2)(8) - 9(5)(-4) - 6(8)(23) =$

$-48 + 576 + 460 - 64 + 180 - 1104 = 1216 - 1216 = \boxed{0}$

Side note: This determinant equals zero because 3 times the top row minus 2 times the middle row equals the bottom row. (One row is a linear combination of the others.)

(12) $\begin{vmatrix} \frac{1}{2} & 12 & \frac{3}{2} \\ 8 & \frac{2}{3} & 4 \\ \frac{1}{4} & 6 & \frac{3}{4} \end{vmatrix} = \frac{1}{2}\left(\frac{2}{3}\right)\left(\frac{3}{4}\right) + 12(4)\left(\frac{1}{4}\right) + \frac{3}{2}(8)(6) - \frac{3}{2}\left(\frac{2}{3}\right)\left(\frac{1}{4}\right) - 12(8)\left(\frac{3}{4}\right) - \frac{1}{2}(4)(6)$

$= \frac{1}{4} + 12 + 72 - \frac{1}{4} - 72 - 12 = \boxed{0}$ Challenge: Can you figure out why this determinant is zero, using the idea from the side note (which applies to rows or columns) to Exercise 11?

5.4 Problems

(1) $x = \boxed{-5}$ and $y = \boxed{8}$

Check: $5(-5) + 4(8) = -25 + 32 = 7$ and $9(-5) + 8(8) = -45 + 64 = 19$.

(2) $x = \boxed{6}$ and $y = \boxed{12}$

Check: $7(6) - 2(12) = 42 - 24 = 18$ and $3(6) + 4(12) = 18 + 48 = 66$.

(3) $x = \boxed{\frac{1}{4}}$ and $y = \boxed{\frac{7}{3}}$

Check: $8\left(\frac{1}{4}\right) - 6\left(\frac{7}{3}\right) = 2 - 14 = -12$ and $4\left(\frac{1}{4}\right) - 9\left(\frac{7}{3}\right) = 1 - 21 = -20$.

(4) $x = \boxed{-\frac{3}{2}}$ and $y = \boxed{\frac{9}{8}}$

Check: $7\left(-\frac{3}{2}\right) + 4\left(\frac{9}{8}\right) = -\frac{21}{2} + \frac{9}{2} = -\frac{12}{2} = -6$ and $6\left(-\frac{3}{2}\right) + 8\left(\frac{9}{8}\right) = -9 + 9 = 0$.

(5) $x = \boxed{3}$ and $y = \boxed{\frac{11}{4}}$

Check: $2(3) + 8\left(\frac{11}{4}\right) = 6 + 22 = 28$ and $9(3) - 4\left(\frac{11}{4}\right) = 27 - 11 = 16$.

(6) $x = \boxed{-\frac{1}{6}}$ and $y = \boxed{-\frac{1}{4}}$

Check: $9\left(-\frac{1}{6}\right) + 2\left(-\frac{1}{4}\right) = -\frac{3}{2} - \frac{1}{2} = -\frac{4}{2} = -2$ and $3\left(-\frac{1}{6}\right) - 6\left(-\frac{1}{4}\right) = -\frac{1}{2} + \frac{3}{2} = \frac{2}{2} = 1$.

(7) $x = \boxed{-2}$, $y = \boxed{7}$, and $z = \boxed{4}$

Check: $8(-2) + 5(7) - 4(4) = -16 + 35 - 16 = 3$, $3(-2) - (7) + 5(4) =$
$-6 - 7 + 20 = 7$, and $-7(-2) - 4(7) + 3(4) = 14 - 28 + 12 = -2$.

(8) $x = \boxed{5}$, $y = \boxed{2}$, and $z = \boxed{9}$

Check: $4(5) - 9(2) = 20 - 18 = 2$, $3(2) + 2(9) = 6 + 18 = 24$, and
$8(5) - 5(9) = 40 - 45 = -5$. Note: $c_1 = a_2 = b_3 = 0$.

(9) $x = \boxed{8}$, $y = \boxed{-3}$, and $z = \boxed{6}$

Check: $6(8) + 7(-3) - 3(6) = 48 - 21 - 18 = 9$, $2(8) - 9(-3) - 7(6) =$
$16 + 27 - 42 = 1$, and $7(8) + 8(-3) + 2(6) = 56 - 24 + 12 = 44$.

(10) $x = \boxed{8}$, $y = \boxed{\frac{1}{2}}$, and $z = \boxed{-1}$

Check: $4(8) - 6\left(\frac{1}{2}\right) + 9(-1) = 32 - 3 - 9 = 20$, $7(8) - 8\left(\frac{1}{2}\right) - 8(-1) =$

$56 - 4 + 8 = 60$, and $3(8) + 4\left(\frac{1}{2}\right) - 6(-1) = 24 + 2 + 6 = 32$.

(11) $x = \boxed{\frac{2}{3}}$, $y = \boxed{\frac{3}{4}}$, and $z = \boxed{\frac{5}{2}}$

Check: $9\left(\frac{2}{3}\right) - 8\left(\frac{3}{4}\right) + 4\left(\frac{5}{2}\right) = 6 - 6 + 10 = 10$, $3\left(\frac{2}{3}\right) + 8\left(\frac{5}{2}\right) = 2 + 20 = 22$, and

$6\left(\frac{2}{3}\right) + 4\left(\frac{3}{4}\right) - 6\left(\frac{5}{2}\right) = 4 + 3 - 15 = -8$. Note: $b_2 = 0$ (which is a coefficient of y).

(12) $x = \boxed{-\frac{1}{3}}$, $y = \boxed{\frac{4}{3}}$, and $z = \boxed{\frac{2}{3}}$

Check: $5\left(-\frac{1}{3}\right) + 7\left(\frac{4}{3}\right) + 2\left(\frac{2}{3}\right) = -\frac{5}{3} + \frac{28}{3} + \frac{4}{3} = \frac{27}{3} = 9$, $8\left(-\frac{1}{3}\right) - 5\left(\frac{4}{3}\right) - 4\left(\frac{2}{3}\right) =$

$-\frac{8}{3} - \frac{20}{3} - \frac{8}{3} = -\frac{36}{3} = -12$, and $4\left(-\frac{1}{3}\right) - 2\left(\frac{4}{3}\right) + 9\left(\frac{2}{3}\right) = -\frac{4}{3} - \frac{8}{3} + 6 = -\frac{12}{3} + 6 = 2$.

5.5 Problems

(1) $x = \boxed{\frac{1}{2}}$ and $y = \boxed{72}$ or $x = \boxed{-\frac{1}{2}}$ and $y = \boxed{-72}$ Check: $3\left(\frac{1}{2}\right)72 = 108$, $\frac{2(72)}{1/2} = \frac{144}{1/2} =$

$144 \div \frac{1}{2} = 144 \times 2 = 288$, $3\left(-\frac{1}{2}\right)(-72) = 108$, and $\frac{2(-72)}{-1/2} = \frac{144}{1/2} = 144 \div \frac{1}{2} = 288$.

(2) $x = \boxed{12}$ and $y = \boxed{15}$ or $x = \boxed{15}$ and $y = \boxed{12}$ Check: $12 + 15 = 27$ and $12(15) = 180$.

(3) $x = \boxed{9}$ and $y = \boxed{18}$ or $x = \boxed{-6}$ and $y = \boxed{3}$ Check: $18 - 9 = 9$, $\frac{1}{9} + \frac{1}{18} = \frac{2}{18} + \frac{1}{18} = \frac{3}{18} = \frac{1}{6}$,

$3 - (-6) = 3 + 6 = 9$, and $\frac{1}{-6} + \frac{1}{3} = -\frac{1}{6} + \frac{2}{6} = \frac{-1+2}{6} = \frac{1}{6}$.

(4) $x = \boxed{24}$ and $y = \boxed{6}$ or $x = \boxed{-18}$ and $y = \boxed{-8}$ Check: $\frac{18}{24} = \frac{3}{4}$ agrees with $\frac{6}{8} = \frac{3}{4}$

and $24 - 6 = 18$ agrees with $3(6) = 18$, and $\frac{18}{-18} = -1$ agrees with $\frac{-8}{8} = -1$ and

$-18 - 6 = -24$ agrees with $3(-8) = -24$.

(5) $x = \boxed{9}$ and $y = \boxed{16}$ Check: $9\sqrt{16} = 9(4) = 36$ and $16\sqrt{9} = 16(3) = 48$.

(6) $x = \boxed{7}$ and $y = \boxed{8}$ or $x = \boxed{-67}$ and $y = \boxed{82}$ Check: $3(7)^2 - 2(8)^2 = 3(49) - 2(64) =$

$147 - 128 = 19$, $7 + 8 = 15$, $3(-67)^2 - 2(82)^2 = 3(4489) - 2(6724) = 13{,}467 - 13{,}448$

$= 19$, and $-67 + 82 = 15$.

(7) $x = \boxed{12}$ and $y = \boxed{25}$ Check: $12 - \sqrt{25} = 12 - 5 = 7$ and $12^2 + 25 = 144 + 25 = 169$.

(8) $x = \boxed{8}$ and $y = \boxed{-3}$ or $x = \boxed{-\frac{432}{65}}$ and $y = \boxed{\frac{349}{65}}$ Check: $4(8) + 7(-3) = 32 - 21 =$

11, $8^2 + (-3)^2 = 64 + 9 = 73$, $4\left(-\frac{432}{65}\right) + 7\left(\frac{349}{65}\right) = -\frac{1728}{65} + \frac{2443}{65} = \frac{715}{65} = 11$, and

$\left(-\frac{432}{65}\right)^2 + \left(\frac{349}{65}\right)^2 = \frac{186{,}624}{4225} + \frac{121{,}801}{4225} = \frac{308{,}425}{4225} = 73$.

6 Polynomials

6.1 Problems

(1) $(5x + 2)(3x^2 - 6x + 4) = 15x^3 - 30x^2 + 20x + 6x^2 - 12x + 8 =$

$\boxed{15x^3 - 24x^2 + 8x + 8}$

(2) $(x - 1)(2x^2 + 8x - 5) = 2x^3 + 8x^2 - 5x - 2x^2 - 8x + 5 = \boxed{2x^3 + 6x^2 - 13x + 5}$

(3) $(x^2 + 6x + 3)(x^2 + 4x + 7) = x^4 + 4x^3 + 7x^2 + 6x^3 + 24x^2 + 42x + 3x^2 + 12x + 21$

$= \boxed{x^4 + 10x^3 + 34x^2 + 54x + 21}$

(4) $(3x^2 + 8x - 5)(4x^2 - 2x + 9) = 12x^4 - 6x^3 + 27x^2 + 32x^3 - 16x^2 + 72x$

$-20x^2 + 10x - 45 = \boxed{12x^4 + 26x^3 - 9x^2 + 82x - 45}$

(5) $(2x^2 - 5x + 6)(5x^2 + 7x - 4) = 10x^4 + 14x^3 - 8x^2 - 25x^3 - 35x^2 + 20x +$

$30x^2 + 42x - 24 = \boxed{10x^4 - 11x^3 - 13x^2 + 62x - 24}$

(6) $(6x^2 - x - 8)^2 = (6x^2 - x - 8)(6x^2 - x - 8) = 36x^4 - 6x^3 - 48x^2 - 6x^3 +$

$x^2 + 8x - 48x^2 + 8x + 64 = \boxed{36x^4 - 12x^3 - 95x^2 + 16x + 64}$

(7) $(x + 3)(4x^2 + 9x - 7) = 4x^3 + 9x^2 - 7x + 12x^2 + 27x - 21 = \boxed{4x^3 + 21x^2 + 20x - 21}$

(8) $(2x - 5)(3x^3 - 8x^2 + 5x - 2) = 6x^4 - 16x^3 + 10x^2 - 4x - 15x^3 + 40x^2 - 25x + 10$

$= \boxed{6x^4 - 31x^3 + 50x^2 - 29x + 10}$

(9) $(3x^2 + 4)(2x^3 + 6x^2 - 7x + 9) = 6x^5 + 18x^4 - 21x^3 + 27x^2 + 8x^3 + 24x^2 - 28x + 36$

$= \boxed{6x^5 + 18x^4 - 13x^3 + 51x^2 - 28x + 36}$

(10) $(8x^2 - 7x)(5x^6 - 4x^4 - 8x^2 + 3) = 40x^8 - 32x^6 - 64x^4 + 24x^2 - 35x^7 +$

$28x^5 + 56x^3 - 21x = \boxed{40x^8 - 35x^7 - 32x^6 + 28x^5 - 64x^4 + 56x^3 + 24x^2 - 21x}$

6.2 Problems

(1) $\boxed{x^2 + 6x + 5}$ Check: $(x + 4)(x^2 + 6x + 5) = x^3 + 6x^2 + 5x + 4x^2 + 24x + 20 =$
$x^3 + 10x^2 + 29x + 20$.

(2) $\boxed{4x^2 + 3x - 7}$ Check: $(x - 6)(4x^2 + 3x - 7) = 4x^3 + 3x^2 - 7x - 24x^2 - 18x + 42 =$
$4x^3 - 21x^2 - 25x + 42$.

(3) $\boxed{6x^4 - 9x^2 + 5}$ Check: $(3x - 8)(6x^4 - 9x^2 + 5) = 18x^5 - 27x^3 + 15x - 48x^4 + 72x^2 - 40 = 18x^5 - 48x^4 - 27x^3 + 72x^2 + 15x - 40.$

(4) $\boxed{2x^3 + 8x^2 - 7x - 12}$ Check: $(5x + 3)(2x^3 + 8x^2 - 7x - 12) = 10x^4 + 40x^3 -35x^2 - 60x + 6x^3 + 24x^2 - 21x - 36 = 10x^4 + 46x^3 - 11x^2 - 81x - 36.$

(5) $\boxed{x^2 + 8x + 3}$ Check: $(4x^2 + 7)(x^2 + 8x + 3) = 4x^4 + 32x^3 + 12x^2 + 7x^2 + 56x + 21$ $= 4x^4 + 32x^3 + 19x^2 + 56x + 21.$ **Tip: Write the divisor as $4x^2 + 0x + 7$.**

(6) $\boxed{4x^2 - 2x + 9}$ Check: $(x^2 + 5x + 4)(4x^2 - 2x + 9) = 4x^4 - 2x^3 + 9x^2 + 20x^3 -10x^2 + 45x + 16x^2 - 8x + 36 = 4x^4 + 18x^3 + 15x^2 + 37x + 36.$

(7) $\boxed{3x^2 + 4x - 7}$ Check: $(2x^2 + 6x - 5)(3x^2 + 4x - 7) = 6x^4 + 8x^3 - 14x^2 + 18x^3 + 24x^2 - 42x - 15x^2 - 20x + 35 = 6x^4 + 26x^3 - 5x^2 - 62x + 35.$

(8) $\boxed{2x^3 + 6x^2 - 12x + 5}$ Check: $(4x^2 - 9x + 8)(2x^3 + 6x^2 - 12x + 5) = 8x^5 + 24x^4 -48x^3 + 20x^2 - 18x^4 - 54x^3 + 108x^2 - 45x + 16x^3 + 48x^2 - 96x + 40 = 8x^5 + 6x^4 -86x^3 + 176x^2 - 141x + 40.$

6.3 Problems

(1) $\boxed{x^2 + 5x + 8 - \dfrac{23}{x+6}}$ Check: $(x + 6)\left(x^2 + 5x + 8 - \dfrac{23}{x+6}\right) = x^3 + 5x^2 + 8x + 6x^2 + 30x + 48 - 23 = x^3 + 11x^2 + 38x + 25.$

(2) $\boxed{6x^2 + 2x - 5 + \dfrac{15}{4x-7}}$ Check: $(4x - 7)\left(6x^2 + 2x - 5 + \dfrac{15}{4x-7}\right) = 24x^3 + 8x^2 - 20x -42x^2 - 14x + 35 + 15 = 24x^3 - 34x^2 - 34x + 50.$

(3) $\boxed{3x^3 - 7x^2 + 5x + 8 + \dfrac{32}{6x-5}}$ Check: $(6x - 5)\left(3x^3 - 7x^2 + 5x + 8 + \dfrac{32}{6x-5}\right) = 18x^4 -42x^3 + 30x^2 + 48x - 15x^3 + 35x^2 - 25x - 40 + 32 = 18x^4 - 57x^3 + 65x^2 + 23x - 8.$

(4) $\boxed{7x^3 + 4x^2 - 9x - 2 - \dfrac{24}{3x+8}}$ Check: $(3x + 8)\left(7x^3 + 4x^2 - 9x - 2 - \dfrac{24}{3x+8}\right) = 21x^4 + 12x^3 - 27x^2 - 6x + 56x^3 + 32x^2 - 72x - 16 - 24 = 21x^4 + 68x^3 + 5x^2 - 78x - 40.$

(5) $\boxed{4x^3 + 9x^2 - 2x + \dfrac{8x}{9x^2+4}}$ Check: $(9x^2 + 4)\left(4x^3 + 9x^2 - 2x + \dfrac{8x}{9x^2+4}\right) = 36x^5 + 81x^4 -18x^3 + 16x^3 + 36x^2 - 8x + 8x = 36x^5 + 81x^4 - 2x^3 + 36x^2.$

(6) $\boxed{6x^2 + 8x - 4 + \dfrac{-5x+9}{5x^2-2x}}$ Check: $(5x^2 - 2x)\left(6x^2 + 8x - 4 + \dfrac{-5x+9}{5x^2-2x}\right) = 30x^4 + 40x^3 -20x^2 - 12x^3 - 16x^2 + 8x - 5x + 9 = 30x^4 + 28x^3 - 36x^2 + 3x + 9.$

(7) $\boxed{5x^3 - 7x^2 + 3x - 6 - \frac{21}{x^2+4x-9}}$ Check: $(x^2 + 4x - 9)\left(5x^3 - 7x^2 + 3x - 6 - \frac{21}{x^2+4x-9}\right) =$

$5x^5 - 7x^4 + 3x^3 - 6x^2 + 20x^4 - 28x^3 + 12x^2 - 24x - 45x^3 + 63x^2 - 27x + 54 - 21) =$

$5x^5 + 13x^4 - 70x^3 + 69x^2 - 51x + 33.$

(8) $\boxed{3x^3 + 9x^2 - 7x + 8 + \frac{4x+11}{2x^2-6x+3}}$ Check: $(2x^2 - 6x + 3)\left(3x^3 + 9x^2 - 7x + 8 + \frac{4x+11}{2x^2-6x+3}\right)$

$= 6x^5 + 18x^4 - 14x^3 + 16x^2 - 18x^4 - 54x^3 + 42x^2 - 48x + 9x^3 + 27x^2 - 21x + 24 +$

$4x + 11 = 6x^5 - 59x^3 + 85x^2 - 65x + 35.$ **Tip: Write $0x^4$ in the dividend.**

Note: There is no x^4 term in the dividend $(6x^5 - 59x^3 + 85x^2 - 65x + 35)$. The x^4 term

cancels out when $2x^2 - 6x + 3$ is multiplied by $3x^3 + 9x^2 - 7x + 8 + \frac{4x+11}{2x^2-6x+3}$.

6.4 Problems

(1) $\boxed{x^2 + 9x - 7 + \frac{2}{x-3}}$ Check: $(x - 3)\left(x^2 + 9x - 7 + \frac{2}{x-3}\right) = x^3 + 9x^2 - 7x - 3x^2 - 27x +$

$21 + 2 = x^3 + 6x^2 - 34x + 23.$

(2) $\boxed{7x^2 + 5x + 4}$ Check: $(x - 8)(7x^2 + 5x + 4) = 7x^3 + 5x^2 + 4x - 56x^2 - 40x - 32 =$

$7x^3 - 51x^2 - 36x - 32.$

(3) $\boxed{6x^2 - 4x + 9 - \frac{1}{x+5}}$ Check: $(x + 5)\left(6x^2 - 4x + 9 - \frac{1}{x+5}\right) = 6x^3 - 4x^2 + 9x + 30x^2$

$-20x + 45 - 1 = 6x^3 + 26x^2 - 11x + 44.$

(4) $\boxed{3x^3 + x^2 - 8x - 5 + \frac{7}{x-6}}$ Check: $(x - 6)\left(3x^3 + x^2 - 8x - 5 + \frac{7}{x-6}\right) = 3x^4 + x^3$

$-8x^2 - 5x - 18x^3 - 6x^2 + 48x + 30 + 7 = 3x^4 - 17x^3 - 14x^2 + 43x + 37.$

(5) $\boxed{5x^3 - 7x^2 + 2x + \frac{5}{x-9}}$ Check: $(x - 9)\left(5x^3 - 7x^2 + 2x + \frac{5}{x-9}\right) = 5x^4 - 7x^3 + 2x^2$

$-45x^3 + 63x^2 - 18x + 5 = 5x^4 - 52x^3 + 65x^2 - 18x + 5.$ Note: There is no constant

term in the answer $\left(5x^3 - 7x^2 + 2x + \frac{5}{x-9}\right)$. There is a zero for this term in the table.

(6) $\boxed{8x^3 + 6x^2 - 9x + 4}$ Check: $(x + 2)(8x^3 + 6x^2 - 9x + 4) = 8x^4 + 6x^3 - 9x^2 + 4x +$

$16x^3 + 12x^2 - 18x + 8 = 8x^4 + 22x^3 + 3x^2 - 14x + 8.$

(7) $\boxed{2x^4 + 5x^3 - 3x^2 + x - 7 - \frac{4}{x+7}}$ Check: $(x + 7)\left(2x^4 + 5x^3 - 3x^2 + x - 7 - \frac{4}{x+7}\right) =$

$2x^5 + 5x^4 - 3x^3 + x^2 - 7x + 14x^4 + 35x^3 - 21x^2 + 7x - 49 - 4 = 2x^5 + 19x^4 + 32x^3$

$-20x^2 - 53$. Note: There is no x term in the dividend $(2x^5 + 19x^4 + 32x^3 - 20x^2 - 53)$. Write a zero for this spot in the table (to the right of the -20 and to the left of the -53).

(8) $\boxed{9x^4 - 7x^3 + 9x - 4}$ Check: $(x - 4)(9x^4 - 7x^3 + 9x - 4) = 9x^5 - 7x^4 + 9x^2 - 4x$ $-36x^4 + 28x^3 - 36x + 16 = 9x^5 - 43x^4 + 28x^3 + 9x^2 - 40x + 16$. Note: There is no x^2 term in the quotient $(9x^4 - 7x^3 + 9x - 4)$. There is a zero for this term in the table.

6.5 Problems

(1) $\boxed{7x + 4 + \dfrac{5}{2x-6}} = \boxed{7x + 4 + \dfrac{5}{2(x-3)}}$ Check: $(2x - 6)\left(7x + 4 + \dfrac{5}{2x-6}\right)$

$= 14x^2 + 8x - 42x - 24 + 5 = 14x^2 - 34x - 19$.

(2) $\boxed{2x^2 - 3x + 8}$ Check: $(4x - 8)(2x^2 - 3x + 8) = 8x^3 - 12x^2 + 32x - 16x^2 + 24x - 64$

$= 8x^3 - 28x^2 + 56x - 64$.

(3) $\boxed{4x^2 + x - 6 - \dfrac{4}{5x+3}}$ Check: $(5x + 3)\left(4x^2 + x - 6 - \dfrac{4}{5x+3}\right) =$

$20x^3 + 5x^2 - 30x + 12x^2 + 3x - 18 - 4 = 20x^3 + 17x^2 - 27x - 22$.

(4) $\boxed{3x^2 + 7 + \dfrac{2}{6x-9}}$ Check: $(6x - 9)\left(3x^2 + 7 + \dfrac{2}{6x-9}\right) = 18x^3 + 42x - 27x^2 - 63 + 2 =$

$18x^3 - 27x^2 + 42x - 61$. Note: There is no x term in the answer $\left(3x^2 + 7 + \dfrac{2}{6x-9}\right)$.

There is a zero for this term in the table.

(5) $\boxed{8x^3 - 5x^2 - 9x + 4}$ Check: $(7x - 4)(8x^3 - 5x^2 - 9x + 4) = 56x^4 - 35x^3 - 63x^2 +$

$28x - 32x^3 + 20x^2 + 36x - 16 = 56x^4 - 67x^3 - 43x^2 + 64x - 16$.

(6) $\boxed{5x^3 + 8x^2 - 2x + 3 - \dfrac{1}{8x+2}}$ Check: $(8x + 2)\left(5x^3 + 8x^2 - 2x + 3 - \dfrac{1}{8x+2}\right) =$

$40x^4 + 64x^3 - 16x^2 + 24x + 10x^3 + 16x^2 - 4x + 6 - 1 = 40x^4 + 74x^3 + 20x + 5$.

Note: There is no x^2 term in the dividend $(40x^4 + 74x^3 + 20x + 5)$. Write a zero for this spot in the table (to the right of the 74 and to the left of the 20).

(7) $x^4 + 8x^3 - 4x^2 + 3x - 6 + \dfrac{3}{9x-6} = \boxed{x^4 + 8x^3 - 4x^2 + 3x - 6 + \dfrac{1}{3x-2}}$ Check:

$(9x - 6)\left(x^4 + 8x^3 - 4x^2 + 3x - 6 + \dfrac{3}{9x-6}\right) = 9x^5 + 72x^4 - 36x^3 + 27x^2 - 54x - 6x^4$

$-48x^3 + 24x^2 - 18x + 36 + 3 = 9x^5 + 66x^4 - 84x^3 + 51x^2 - 72x + 39$.

Note: $\dfrac{3}{9x-6} = \dfrac{3}{3(3x-2)} = \dfrac{1}{3x-2}$. We reduced the answer to express it in standard form.

(8) $\boxed{6x^3 - 8x^2 + 9x - 7}$ Check: $(3-x)(6x^3 - 8x^2 + 9x - 7) =$
$18x^3 - 24x^2 + 27x - 21 - 6x^4 + 8x^3 - 9x^2 + 7x = -6x^4 + 26x^3 - 33x^2 + 34x - 21.$

6.6 Problems

(1) $p(4) = 5(4) - 7 = 20 - 7 = \boxed{13}$ Note: $c = 4$. Compare $x - 4$ to $x - c$.

(2) $p(7) = (7)^2 - 9(7) + 14 = 49 - 63 + 14 = \boxed{0}$ Note: $c = 7$. Compare $x - 4$ to $x - c$.

(3) $p(-6) = 2(-6)^3 + 8(-6)^2 - 7(-6) + 52 = 2(-216) + 8(36) + 42 + 52 =$
$-432 + 288 + 94 = \boxed{-50}$ Note: $c = -6$. Compare $x + 6$ to $x - c$.

(4) $p(-1) = 5(-1)^5 - 3(-1)^4 + 6(-1)^3 - 7(-1)^2 + 9(-1) - 15 = 5(-1) - 3(1) + 6(-1)$
$-7(1) - 9 - 15 = -5 - 3 - 6 - 7 - 24 = \boxed{-45}$ Note: $c = -1$. Compare $x + 1$ to $x - c$.

(5) $p\left(\frac{1}{4}\right) = 8\left(\frac{1}{4}\right)^2 + 6\left(\frac{1}{4}\right) - 5 = \frac{8}{16} + \frac{6}{4} - 5 = \frac{1}{2} + \frac{3}{2} - 5 = \frac{4}{2} - 5 = 2 - 5 = \boxed{-3}$ Note:
$a = 4$ and $b = 1$. $\frac{b}{a} = \frac{1}{4}$. Compare $4x - 1$ to $ax - b$.

(6) $p\left(-\frac{3}{2}\right) = 6\left(-\frac{3}{2}\right)^3 + 4\left(-\frac{3}{2}\right)^2 - 3\left(-\frac{3}{2}\right) + 7 = 6\left(-\frac{27}{8}\right) + 4\left(\frac{9}{4}\right) + \frac{9}{2} + 7 = -\frac{81}{4} + \frac{36}{4} +$
$\frac{18}{4} + \frac{28}{4} = \boxed{\frac{1}{4}} = \boxed{0.25}$ Note: $a = 2$ and $b = -3$. $\frac{b}{a} = -\frac{3}{2}$. Compare $2x + 3$ to $ax - b$.

(7) $p\left(\frac{5}{2}\right) = 16\left(\frac{5}{2}\right)^5 + 64\left(\frac{5}{2}\right)^4 - 80\left(\frac{5}{2}\right)^3 + 48\left(\frac{5}{2}\right)^2 - 72\left(\frac{5}{2}\right) + 39 = 16\left(\frac{3125}{32}\right) + 64\left(\frac{625}{16}\right)$
$-80\left(\frac{125}{8}\right) + 48\left(\frac{25}{4}\right) - 180 + 39 = \frac{3125}{2} + 2500 - 1250 + 300 - 141 = \frac{3125}{2} + 1409 =$
$\boxed{\frac{5943}{2}} = \boxed{2971.5}$ Note: $a = 2$ and $b = 5$. $\frac{b}{a} = \frac{5}{2}$. Compare $2x - 5$ to $ax - b$.

(8) $p\left(\frac{1}{2}\right) = 4\left(\frac{1}{2}\right)^4 = \frac{4}{16} = \boxed{\frac{1}{4}} = \boxed{0.25}$ Note: $a = 6$ and $b = 3$. $\frac{b}{a} = \frac{3}{6} = \frac{1}{2}$. Compare
$6x - 3$ to $ax - b$.

(9) $p(-2) = (-2)^5 - 3(-2)^4 - 5(-2)^3 + 6(-2)^2 - 7(-2) + 10 = -32 - 3(16)$
$-5(-8) + 6(4) + 14 + 10 = -32 - 48 + 40 + 24 + 14 + 10 = \boxed{8}$ Note: $a = 4$ and $b =$
-8. $\frac{b}{a} = \frac{-8}{4} = -2$. Compare $4x + 8$ to $ax - b$.

(10) $\frac{8x^3 + 3x + 19}{4x + 5}$ $p\left(-\frac{5}{4}\right) = 8\left(-\frac{5}{4}\right)^3 + 3\left(-\frac{5}{4}\right) + 19 = 8\left(-\frac{125}{64}\right) - \frac{15}{4} + 19 = -\frac{125}{8} - \frac{30}{8} + \frac{152}{8} =$
$\boxed{-\frac{3}{8}} = \boxed{-0.375}$ Note: $a = 4$ and $b = -5$. $\frac{b}{a} = -\frac{5}{4}$. Compare $4x + 5$ to $ax - b$.

6.7 Problems

(1) $x = 3$, $x = 5$, and $x = 7$. Check: $p(3) = 3^3 - 15(3)^2 + 71(3) - 105 = 27 - 135 + 213 - 105 = 0$, $p(5) = 5^3 - 15(5)^2 + 71(5) - 105 = 125 - 375 + 355 - 105 = 0$, and $p(7) = 7^3 - 15(7)^2 + 71(7) - 105 = 343 - 735 + 497 - 105 = 0$.

$$(x - 3)(x - 5)(x - 7) = (x^2 - 8x + 15)(x - 7)$$
$$= x^3 - 8x^2 + 15x - 7x^2 + 56x - 105 = x^3 - 15x^2 + 71x - 105$$

(2) $x = -15$, $x = -5$, and $x = -2$. Check: $p(-15) = 4(-15)^3 + 88(-15)^2 + 460(-15) + 600 = -13,500 + 19,800 - 6900 + 600 = 0$, $p(-5) = 4(-5)^3 + 88(-5)^2 + 460(-5) + 600 = -500 + 2200 - 2300 + 600 = 0$, and $p(-2) = 4(-2)^3 + 88(-2)^2 + 460(-2) + 600 = -32 + 352 - 920 + 600 = 0$.

Tip: This problem is simpler if you factor the 4 out in the beginning.

$$4(x + 2)(x + 5)(x + 15) = 4(x^2 + 7x + 10)(x + 15)$$
$$= 4(x^3 + 7x^2 + 10x + 15x^2 + 105x + 150) = 4(x^3 + 22x^2 + 115x + 150)$$
$$= 4x^3 + 88x^2 + 460x + 600$$

(3) $x = -6$, $x = -1$, and $x = 4$. Check: $p(-6) = (-6)^3 + 3(-6)^2 - 22(-6) - 24 = -216 + 108 + 132 - 24 = 0$, $p(-1) = (-1)^3 + 3(-1)^2 - 22(-1) - 24 = -1 + 3 + 22 - 24 = 0$, and $p(4) = (4)^3 + 3(4)^2 - 22(4) - 24 = 64 + 48 - 88 - 24 = 0$.

$$(x + 1)(x - 4)(x + 6) = (x^2 - 3x - 4)(x + 6)$$
$$= x^3 - 3x^2 - 4x + 6x^2 - 18x - 24 = x^3 + 3x^2 - 22x - 24$$

(4) $x = -9$, $x = -4$, $x = 2$, and $x = 5$. Check: $p(-9) = (-9)^4 + 6(-9)^3 - 45(-9)^2 - 122(-9) + 360 = 6561 - 4374 - 3645 + 1098 + 360 = 0$, $p(-4) = (-4)^4 + 6(-4)^3 - 45(-4)^2 - 122(-4) + 360 = 256 - 384 - 720 + 488 + 360 = 0$, $p(2) = 2^4 + 6(2)^3 - 45(2)^2 - 122(2) + 360 = 16 + 48 - 180 - 244 + 360 = 0$, and $p(5) = 5^4 + 6(5)^3 - 45(5)^2 - 122(5) + 360 = 625 + 750 - 1125 - 610 + 360 = 0$.

$$(x - 2)(x + 4)(x - 5)(x + 9) = (x^2 + 2x - 8)(x^2 + 4x - 45)$$
$$= x^4 + 2x^3 - 8x^2 + 4x^3 + 8x^2 - 32x - 45x^2 - 90x + 360$$
$$= x^4 + 6x^3 - 45x^2 - 122x + 360$$

(5) $x = 3$, $x = 3$, $x = 5$, and $x = 7$. Note: $x = 3$ is a double root. Check: $p(3) = 3^4 - 18(3)^3 + 116(3)^2 - 318(3) + 315 = 81 - 486 + 1044 - 954 + 315 = 0$, $p(5) = 5^4 - 18(5)^3 + 116(5)^2 - 318(5) + 315 = 625 - 2250 + 2900 - 1590 + 315 = 0$, and $p(7) = 7^4 - 18(7)^3 + 116(7)^2 - 318x + 315 = 2401 - 6174 + 5684 - 2226 + 315 = 0$.

$$(x - 3)(x - 3)(x - 5)(x - 7) = (x^2 - 6x + 9)(x^2 - 12x + 35)$$
$$= x^4 - 6x^3 + 9x^2 - 12x^3 + 72x^2 - 108x + 35x^2 - 210x + 315$$
$$= x^4 - 18x^3 + 116x^2 - 318x + 315$$

(6) $x = -6$, $x = 0$, $x = 6$, and $x = 8$. Check: $p(-6) = (-6)^4 - 8(-6)^3 - 36(-6)^2 + 288(-6) = 1296 + 1728 - 1296 - 1728 = 0$, $p(0) = 0$, $p(6) = 6^4 - 8(6)^3 - 36(6)^2 + 288(6) = 1296 - 1728 - 1296 + 1728 = 0$, and $p(8) = 8^4 - 8(8)^3 - 36(8)^2 + 288(8) = 4096 - 4096 - 2304 + 2304 = 0$.

$$x(x - 6)(x + 6)(x - 8) = (x^3 - 36x)(x - 8) = x^4 - 8x^3 - 36x^2 + 288x$$

(7) $x = -\frac{1}{2}$, $x = \frac{3}{5}$, and $x = \frac{2}{3}$. Check: $p\left(-\frac{1}{2}\right) = 30\left(-\frac{1}{2}\right)^3 - 23\left(-\frac{1}{2}\right)^2 - 7\left(-\frac{1}{2}\right) + 6 = -\frac{15}{4} - \frac{23}{4} + \frac{14}{4} + \frac{24}{4} = 0$, $p\left(\frac{3}{5}\right) = 30\left(\frac{3}{5}\right)^3 - 23\left(\frac{3}{5}\right)^2 - 7\left(\frac{3}{5}\right) + 6 = \frac{162}{25} - \frac{207}{25} - \frac{105}{25} + \frac{150}{25} = 0$, and $p\left(\frac{2}{3}\right) = 30\left(\frac{2}{3}\right)^3 - 23\left(\frac{2}{3}\right)^2 - 7\left(\frac{2}{3}\right) + 6 = \frac{80}{9} - \frac{92}{9} - \frac{42}{9} + \frac{54}{9} = 0$.

$$(2x + 1)(3x - 2)(5x - 3) = (6x^2 - x - 2)(5x - 3)$$
$$= 30x^3 - 5x^2 - 10x - 18x^2 + 3x + 6 = 30x^3 - 23x^2 - 7x + 6$$

(8) $x = -\frac{8}{5}$, $x = \frac{4}{3}$, and $x = 3$. Check: $p\left(-\frac{8}{5}\right) = 15\left(-\frac{8}{5}\right)^3 - 41\left(-\frac{8}{5}\right)^2 - 44\left(-\frac{8}{5}\right) + 96 = -\frac{1536}{25} - \frac{2624}{25} + \frac{1760}{25} + \frac{2400}{25} = 0$, $p\left(\frac{4}{3}\right) = 15\left(\frac{4}{3}\right)^3 - 41\left(\frac{4}{3}\right)^2 - 44\left(\frac{4}{3}\right) + 96 = \frac{320}{9} - \frac{656}{9} - \frac{528}{9} + \frac{864}{9} = 0$, and $p(3) = 15(3)^3 - 41(3)^2 - 44(3) + 96 = 405 - 369 - 132 + 96 = 0$.

$$(3x - 4)(5x + 8)(x - 3) = (15x^2 + 4x - 32)(x - 3)$$
$$= 15x^3 + 4x^2 - 32x - 45x^2 - 12x + 96 = 15x^3 - 41x^2 - 44x + 96$$

(9) $x = -\frac{9}{4}$, $x = -1$, and $x = \frac{8}{3}$. Check: $p\left(-\frac{9}{4}\right) = 12\left(-\frac{9}{4}\right)^3 + 7\left(-\frac{9}{4}\right)^2 - 77\left(-\frac{9}{4}\right) - 72 = -\frac{2187}{16} + \frac{567}{16} + \frac{2772}{16} - \frac{1152}{16} = 0$, $p(-1) = 12(-1)^3 + 7(-1)^2 - 77(-1) - 72 = -12 + 7 + 77 - 72 = 0$, and $p\left(\frac{8}{3}\right) = 12\left(\frac{8}{3}\right)^3 + 7\left(\frac{8}{3}\right)^2 - 77\left(\frac{8}{3}\right) - 72 = \frac{2048}{9} + \frac{448}{9} - \frac{1848}{9} - \frac{648}{9} = 0$.

$$(x + 1)(3x - 8)(4x + 9) = (x + 1)(12x^2 - 5x - 72)$$
$$= 12x^3 - 5x^2 - 72x + 12x^2 - 5x - 72 = 12x^3 + 7x^2 - 77x - 72$$

(10) $x = 5$, $x = -3i$, and $x = 3i$. Check: $p(5) = 6(5)^3 - 30(5)^2 + 54(5) - 270 = 750 - 750 + 270 - 270 = 0$, $p(-3i) = 6(-3i)^3 - 30(-3i)^2 + 54(-3i) - 270 = 6(-3)^3(i)^3 - 30(-3)^2(i)^2 - 162i - 270 = -162(-i) - 270(-1) - 162i - 270 = 162i + 270 - 162i - 270 = 0$, and $p(3i) = 6(3i)^3 - 30(3i)^2 + 54(3i) - 270 = 162(i)^3 - 270(i)^2 + 162i - 270 = -162i + 270 + 162i - 270 = 0$. Note: Recall from Sec. 3.4 that $i^3 = -i$. Tip: This

problem is simpler if you factor the 6 out in the beginning.

$$6(x-5)(x-3i)(x+3i) = (6x-30)(x^2+9) = 6x^3 - 30x^2 + 54x - 270$$

(11) $x = -9$, $x = -2\sqrt{3}$, and $x = 2\sqrt{3}$. Check: $p(-9) = 3(-9)^3 + 27(-9)^2 - 36(-9) -$

$324 = -2187 + 2187 + 324 - 324 = 0$, $p(-2\sqrt{3}) = 3(-2\sqrt{3})^3 + 27(-2\sqrt{3})^2 -$

$36(-2\sqrt{3}) - 324 = 3(-2)^3(\sqrt{3})^3 + 27(-2)^2(\sqrt{3})^2 + 72\sqrt{3} - 324 = -3(8)(3\sqrt{3}) +$

$27(4)(3) + 72\sqrt{3} - 324 = -72\sqrt{3} + 324 + 72\sqrt{3} - 324 = 0$, and $p(2\sqrt{3}) = 3(2\sqrt{3})^3 +$

$27(2\sqrt{3})^2 - 36(2\sqrt{3}) - 324 = 3(2)^3(\sqrt{3})^3 + 27(2)^2(\sqrt{3})^2 - 72\sqrt{3} - 324 = 3(8)(3\sqrt{3}) +$

$27(4)(3) - 72\sqrt{3} - 324 = 72\sqrt{3} + 324 - 72\sqrt{3} - 324 = 0$. Tip: This problem is simpler if you factor the 3 out in the beginning.

$$3(x+9)(x-2\sqrt{3})(x+2\sqrt{3}) = (3x+27)(x^2-12) = 3x^3 + 27x^2 - 36x - 324$$

(12) $x = 0$, $x = 0$, $x = \frac{2}{3}$, $x = -i\sqrt{2}$, and $x = i\sqrt{2}$. Note: $x = 0$ is a double root. Check:

$$p(0) = 0, p\left(\frac{2}{3}\right) = 15\left(\frac{2}{3}\right)^5 - 10\left(\frac{2}{3}\right)^4 + 30\left(\frac{2}{3}\right)^3 - 20\left(\frac{2}{3}\right)^2 = \frac{160}{81} - \frac{160}{81} + \frac{80}{9} - \frac{80}{9} = 0,$$

$p(-i\sqrt{2}) = 15(-i\sqrt{2})^5 - 10(-i\sqrt{2})^4 + 30(-i\sqrt{2})^3 - 20(-i\sqrt{2})^2 = 15(i)^5(-\sqrt{2})^5 -$

$10(i)^4(-\sqrt{2})^4 + 30(i)^3(-\sqrt{2})^3 - 20(i)^2(-\sqrt{2})^2 = 15i(2)^2(-\sqrt{2}) - 10(1)(2)^2 +$

$30(-i)(2)(-\sqrt{2}) - 20(-1)(2) = -60i\sqrt{2} - 40 + 60i\sqrt{2} + 40 = 0$, and $p(i\sqrt{2}) =$

$15(i\sqrt{2})^5 - 10(i\sqrt{2})^4 + 30(i\sqrt{2})^3 - 20(i\sqrt{2})^2 = 15(i)^5(\sqrt{2})^5 - 10(i)^4(\sqrt{2})^4 +$

$30(i)^3(\sqrt{2})^3 - 20(i)^2(\sqrt{2})^2 = 15i(2)^2(\sqrt{2}) - 10(1)(2)^2 + 30(-i)(2)(\sqrt{2}) - 20(-1)(2) =$

$60i\sqrt{2} - 40 - 60i\sqrt{2} + 40 = 0$. Tip: Factor out $5x^2$.

$$5x^2(3x-2)(x+i\sqrt{2})(x-i\sqrt{2}) = (15x^3 - 10x^2)(x^2+2)$$
$$= 15x^5 - 10x^4 + 30x^3 - 20x^2$$

7 Graphs

7.1 Problems

(1) pass

(2) pass

(3) pass

(4) pass

(5) fail

(6) fail

(7) pass

(8) fail

(9) pass

(10) pass

(11) pass

(12) pass

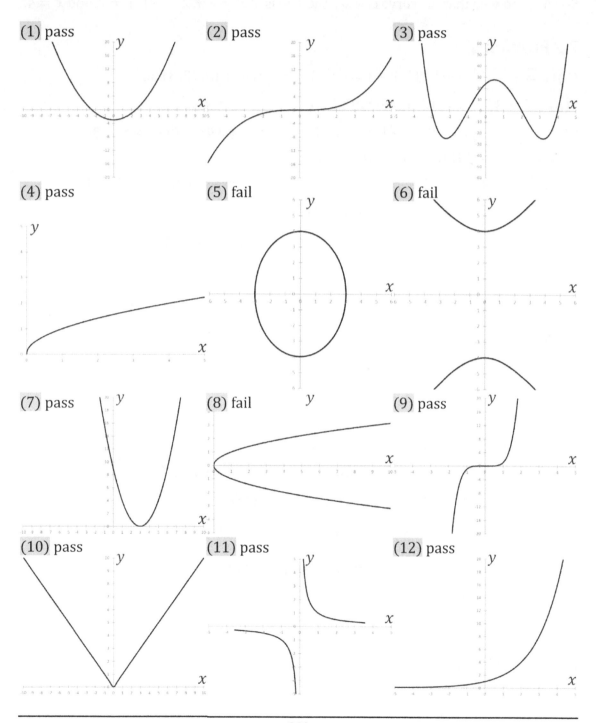

Notes: If a graph passed the vertical line test, y is a function of x. If a graph failed the vertical line test, y is not a function of x. The equation in Exercise 3 can be factored as $x^4 - 2x^3 - 13x^2 + 14x + 24 = (x - 4)(x - 2)(x + 1)(x + 3)$ using the method from Sec. 6.7, showing that the curve crosses the x-axis at $x = -3, x = -1, x = 2$, and $x = 4$.

7.2 Problems

(1) $\boxed{y = 3x - 3}$ Check: $3(2) - 3 = 6 - 3 = 3$ agrees with $(2, 3)$ and $3(5) - 3 = 15 - 3 = 12$ agrees with $(5, 12)$.

(2) $\boxed{y = 2x - 2}$ Check: $2(-2) - 2 = -4 - 2 = -6$ agrees with $(-2, -6)$ and $2(1) - 2 = 2 - 2 = 0$ agrees with $(1, 0)$.

(3) $\boxed{y = -6x + 3}$ Check: $-6(-1) + 3 = 6 + 3 = 9$ agrees with $(-1, 9)$ and $-6(1) + 3 = -6 + 3 = -3$ agrees with $(1, -3)$.

(4) $\boxed{y = \frac{x}{2} + \frac{7}{2}}$ Alternate answer: $y = \frac{x+7}{2}$. Check: $\frac{3}{2} + \frac{7}{2} = \frac{10}{2} = 5$ agrees with $(3, 5)$ and $\frac{7}{2} + \frac{7}{2} = \frac{14}{2} = 7$ agrees with $(7, 7)$.

(5) $\boxed{y = 3x + 6}$ Check: $3(0) + 6 = 0 + 6 = 6$ agrees with $(0, 6)$ and $3(-5) + 6 = -15 + 6 = -9$ agrees with $(-5, -9)$.

(6) $\boxed{y = -4x - 5}$ Check: $-4(-4) - 5 = 16 - 5 = 11$ agrees with $(-4, 11)$ and $-4(-1) - 5 = 4 - 5 = -1$ agrees with $(-1, -1)$.

(7) $\boxed{a = 6}$ Check: $2(6) + 3 = 12 + 3 = 15$ agrees with $(a, 15)$.

(8) $\boxed{c = 4}$ Check: $-(-4) = 4$ agrees with $(-4, c)$.

(9) $\boxed{d = -11}$ Check: $4(5) - 31 = 20 - 31 = -11$ agrees with $(5, d)$.

(10) $\boxed{e = 8}$ Check: $\frac{8}{2} + 8 = 4 + 8 = 12$ agrees with $(e, 12)$.

(11) $\boxed{f = 9}$ Check: $-\frac{9}{3} - 6 = -3 - 6 = -9$ agrees with $(f, -9)$.

(12) $\boxed{g = -10}$ Check: $\frac{5(-8)}{4} = -\frac{40}{4} = -10$ agrees with $(-8, g)$.

(13) $\boxed{a = 8}$ Check: $m = \frac{12-3}{8-5} = \frac{9}{3} = 3$.

(14) $\boxed{c = -3}$ Check: $m = \frac{-3-4}{5-(-2)} = \frac{-7}{5+2} = \frac{-7}{7} = -1$.

(15) $\boxed{d = -2}$ Check: $m = \frac{15-(-10)}{3-(-2)} = \frac{15+10}{5} = \frac{25}{5} = 5$.

(16) $\boxed{e = 0}$ Check: $m = \frac{4-0}{9-1} = \frac{4}{8} = \frac{1}{2}$.

(17) $\boxed{f = 11}$ Check: $m = \frac{11-11}{5-(-7)} = \frac{0}{5+7} = \frac{0}{12} = 0$.

(18) $\boxed{g = 4}$ Check: $m = \frac{-16-8}{4-0} = \frac{-24}{4} = -6$.

(19) x-intercept $= \boxed{-\frac{4}{3}}$ and y-intercept $= b = \boxed{8}$ Check: $6\left(-\frac{4}{3}\right) + 8 = -\frac{24}{3} + 8 =$ $-8 + 8 = 0$ agrees with $\left(-\frac{4}{3}, 0\right)$ and $6(0) + 8 = 8$ agrees with $(0, 8)$.

(20) x-intercept $= \boxed{0}$ and y-intercept $= b = \boxed{0}$ Check: $-\frac{0}{4} = 0$ agrees with $(0, 0)$.

(21) x-intercept $= \boxed{4}$ and y-intercept $= b = \boxed{12}$ Check: $12 - 3(4) = 12 - 12 = 0$ agrees with $(4, 0)$ and $12 - 3(0) = 12 - 0 = 12$ agrees with $(0, 12)$.

(22) x-intercept $= \boxed{\frac{1}{3}}$ and y-intercept $= b = \boxed{-2}$ Check: $6\left(\frac{1}{3}\right) - 2 = 2 - 2 = 0$ agrees with $\left(\frac{1}{3}, 0\right)$ and $6(0) - 2 = 0 - 2 = -2$ agrees with $(0, -2)$. Note: $\frac{2}{6} = \frac{1}{3}$.

(23) x-intercept $= \boxed{-6}$ and y-intercept $= b = \boxed{-24}$ Check: $-4(-6) - 24 = 24 - 24 = 0$ agrees with $(-6, 0)$ and $-4(0) - 24 = 0 - 24 = -24$ agrees with $(0, -24)$.

(24) x-intercept $= \boxed{1}$ and y-intercept $= b = \boxed{-4}$ Check: Compare $y - 8 = 0 - 8 =$ -8 with $4(x - 3) = 4(1 - 3) = 4(-2) = -8$ to see that this agrees with $(1, 0)$ and compare $y - 8 = -4 - 8 = -12$ with $4(x - 3) = 4(0 - 3) = 4(-3) = -12$ to see that this agrees with $(0, -4)$. Notes: $y - 8 = 4(x - 3)$ is in point-slope form instead of slope-intercept form. Distribute to get $y - 8 = 4x - 12$ and add 8 to both sides to get $y = 4x - 12 + 8 = 4x - 4$. The given equation is equivalent to $y = 4x - 4$.

(25) x-intercept $= \boxed{8}$ and y-intercept $= b = \boxed{16}$ Checks: $m = \frac{0-12}{8-2} = \frac{-12}{6} = -2$ for $(2, 12)$ and $(8, 0)$ agrees with $m = \frac{0-16}{8-0} = \frac{-16}{8} = -2$ for $(0, 16)$ and $(8, 0)$. Also, $y = -2x + 16 = -2(2) + 16 = -4 + 16 = 12$ agrees with $(2, 12)$ and $y = -2x + 16 = -2(8) + 16 = -16 + 16 = 0$ agrees with $(8, 0)$.

(26) x-intercept $= \boxed{9}$ and y-intercept $= b = \boxed{9}$ Checks: $m = \frac{3-6}{6-3} = \frac{-3}{3} - -1$ for $(3, 6)$ and $(6, 3)$ agrees with $m = \frac{0-9}{9-0} = \frac{-9}{9} = -1$ for $(0, 9)$ and $(9, 0)$. Also, $y = -x + 9 = -3 + 9 = 6$ agrees with $(3, 6)$ and $y = -x + 9 = -6 + 9 = 3$ agrees with $(6, 3)$.

(27) x-intercept = $\boxed{4}$ and y-intercept = $b = \boxed{4}$ Checks: $m = \frac{-4-8}{8-(-4)} = \frac{-12}{8+4} = \frac{-12}{12} = -1$

for $(-4, 8)$ and $(8, -4)$ agrees with $m = \frac{0-4}{4-0} = \frac{-4}{4} = -1$ for $(0, 4)$ and $(4, 0)$. Also, $y =$

$-x + 4 = -(-4) + 4 = 4 + 4 = 8$ agrees with $(-4, 8)$ and $y = -x + 4 = -8 + 4 =$

-4 agrees with $(8, -4)$.

(28) x-intercept = $\boxed{\frac{48}{5}} = \boxed{9.6}$ and y-intercept = $b = \boxed{24}$ Checks: $m = \frac{9-24}{6-0} = \frac{-15}{6} = -\frac{5}{2}$

for $(0, 24)$ and $(6, 9)$ agrees with $m = \frac{0-24}{\frac{48}{5}-0} = -24 \div \frac{48}{5} = -24 \times \frac{5}{48} = -\frac{5}{2}$ for $(0, 24)$

and $(9.6, 0)$. Also, $y = -\frac{5}{2}x + 24 = -\frac{5}{2}(0) + 24 = 0 + 24 = 24$ agrees with $(0, 24)$ and

$y = -\frac{5}{2}x + 24 = -\frac{5}{2}(6) + 24 = -\frac{30}{2} + 24 = -15 + 24 = 9$ agrees with $(6, 9)$.

(29) x-intercept = $\boxed{-3}$ and y-intercept = $b = \boxed{3}$ Checks: $m = \frac{5-(-3)}{2-(-6)} = \frac{5+3}{2+6} = \frac{8}{8} = 1$ for

$(-6, -3)$ and $(2, 5)$ agrees with $m = \frac{0-3}{-3-0} = \frac{-3}{-3} = 1$ for $(0, 3)$ and $(-3, 0)$. Also, $y = x + 3 =$

$-6 + 3 = -3$ agrees with $(-6, -3)$ and $y = x + 3 = 2 + 3 = 5$ agrees with $(2, 5)$.

(30) x-intercept = $\boxed{9}$ and y-intercept = $b = \boxed{-\frac{27}{4}}$ Checks: $m = \frac{-6-(-9)}{1-(-3)} = \frac{-6+9}{1+3} = \frac{3}{4}$ for

$(-3, -9)$ and $(1, -6)$ agrees with $m = \frac{0-\left(-\frac{27}{4}\right)}{9-0} = \frac{27/4}{9} = \frac{27}{4} \div 9 = \frac{27}{4} \times \frac{1}{9} = \frac{3}{4}$ for $\left(0, -\frac{27}{4}\right)$

and $(9, 0)$. Also, $y = \frac{3}{4}x - \frac{27}{4} = \frac{3}{4}(-3) - \frac{27}{4} = -\frac{9}{4} - \frac{27}{4} = -\frac{36}{4} = -9$ agrees with $(-3, -9)$

and $y = \frac{3}{4}x - \frac{27}{4} = \frac{3}{4}(1) - \frac{27}{4} = \frac{3}{4} - \frac{27}{4} = -\frac{24}{4} = -6$ agrees with $(1, -6)$.

(31) $\boxed{a = -6}$ Check: $m = \frac{-6-2}{9-3} = -\frac{8}{6} = -\frac{4}{3}$, $y = -\frac{4}{3}x + 6 = -\frac{4}{3}(3) + 6 = -4 + 6 = 2$

agrees with $(3, 2)$, and $y = -\frac{4}{3}x + 6 = -\frac{4}{3}(9) + 6 = -12 + 6 = -6$ agrees with $(9, -6)$.

Tip: First use $(3, 2)$ and $(0, 6)$ to find the slope.

(32) $\boxed{c = 5}$ Check: $m = \frac{11-5}{8-2} = \frac{6}{6} = 1$, $y = x + 3 = 2 + 3 = 5$ agrees with $(2, 5)$, and $y =$

$x + 3 = 8 + 3 = 11$ agrees with $(8, 11)$. Tip: First use $(8, 11)$ and $(0, 3)$ to find the slope.

(33) $\boxed{d = -14}$ Check: $m = \frac{3-18}{-4-(-14)} = \frac{-15}{-4+14} = \frac{-15}{10} = -\frac{3}{2}$, $y = -\frac{3}{2}x - 3 = -\frac{3}{2}(-14) - 3 =$

$21 - 3 = 18$ agrees with $(-14, 18)$, and $y = -\frac{3}{2}x - 3 = -\frac{3}{2}(-4) - 3 = 6 - 3 = 3$ agrees

with $(-4, 3)$. Tip: First use $(-4, 3)$ and $(0, -3)$ to find the slope.

(34) $\boxed{e = -6}$ Check: $m = \frac{15-3}{-6-6} = \frac{12}{-12} = -1$, $y = -x + 9 = -6 + 9 = 3$ agrees with $(6, 3)$, and $y = -x + 9 = -(-6) + 9 = 6 + 9 = 15$ agrees with $(-6, 15)$. Tip: First use $(6, 3)$ and $(9, 0)$ to find the slope. Note: It is just coincidental that the x- and y-intercepts happen to be the same in this problem. You need to **solve** for b before you set $b = 9$.

Note: Exercises 34-36 give you the x-intercept (NOT the y-intercept).

(35) $\boxed{f = 15}$ Check: $m = \frac{5-15}{3-1} = \frac{-10}{2} = -5$, $y = -5x + 20 = -5(1) + 20 = -5 + 20 = 15$ agrees with $(1, 15)$, and $y = -5x + 20 = -5(3) + 20 = -15 + 20 = 5$ agrees with $(3, 5)$. Tip: First use $(3, 5)$ and $(4, 0)$ to find the slope. Note: The x-intercept is 4, whereas the y-intercept is 20.

(36) $\boxed{g = 0}$ Check: $m = \frac{4-10}{-3-0} = \frac{-6}{-3} = 2$, $y = 2x + 10 = 2(0) + 10 = 0 + 10 = 10$ agrees with $(0, 10)$, and $y = 2x + 10 = 2(-3) + 10 = -6 + 10 = 4$ agrees with $(-3, 4)$. Tip: First use $(-3, 4)$ and $(-5, 0)$ to find the slope. Note: The x-intercept is -5, whereas the y-intercept is 10.

(37) $\boxed{y = -x + 4}$ Check: $y = -4 + 4 = 0$ agrees with $(4, 0)$ and $y = -0 + 4 = 4$ agrees with $(0, 4)$. Tip: First use $(4, 0)$ and $(0, 4)$ to find the slope.

(38) $\boxed{y = 2x + 4}$ Check: $y = 2(-2) + 4 = -4 + 4 = 0$ agrees with $(-2, 0)$ and $y = 2(0) + 4 = 0 + 4 = 4$ agrees with $(0, 4)$.

(39) $\boxed{y = 3x - 9}$ Check: $y = 3(3) - 9 = 9 - 9 = 0$ agrees with $(3, 0)$ and $y = 3(0) - 9 = 0 - 9 = -9$ agrees with $(0, -9)$.

(40) $\boxed{y = -\frac{x}{4} + 3}$ Check: $y = -\frac{12}{4} + 3 = -3 + 3 = 0$ agrees with $(12, 0)$ and $y = -\frac{0}{4} + 3 = 0 + 3 = 3$ agrees with $(0, 3)$.

(41) $\boxed{y = \frac{x}{3} + 2}$ Check: $y = \frac{-6}{3} + 2 = -2 + 2 = 0$ agrees with $(-6, 0)$ and $y = \frac{0}{3} + 2 = 0 + 2 = 2$ agrees with $(0, 2)$.

(42) $\boxed{y = -\frac{x}{2} - 5}$ Check: $y = -\frac{(-10)}{2} - 5 = 5 - 5 = 0$ agrees with $(-10, 0)$ and $y = -\frac{0}{2} - 5 = 0 - 5 = -5$ agrees with $(0, -5)$.

(43) $\boxed{(7, 33)}$ Check: $y = 4x + 5 = 4(7) + 5 = 28 + 5 = 33$ and $y = 7x - 16 = 7(7) - 16 = 49 - 16 = 33$.

(44) $\boxed{(-2,-13)}$ Check: $y = 2x - 9 = 2(-2) - 9 = -4 - 9 = -13$ and $y = 6x - 1 = 6(-2) - 1 = -12 - 1 = -13$.

(45) $\boxed{(-66,-30)}$ Check: $y = \frac{x}{3} - 8 = \frac{-66}{3} - 8 = -22 - 8 = -30$ and $y = \frac{x}{2} + 3 = \frac{-66}{2} + 3 = -33 + 3 = -30$.

(46) $\boxed{(4,5)}$ Check: $y = 5x - 15 = 5(4) - 15 = 20 - 15 = 5$ and $y = 9 - x = 9 - 4 = 5$.

(47) $\boxed{(-3,17)}$ Check: $y = -3x + 8 = -3(-3) + 8 = 9 + 8 = 17$.

(48) $\boxed{(-9,-61)}$ Check: $y - 3 = -61 - 3 = -64$ agrees with $4(x - 7) = 4(-9 - 7) = 4(-16) = -64$ and $y = 9x + 20 = 9(-9) + 20 = -81 + 20 = -61$. Notes: $y - 3 = 4(x - 7)$ is in point-slope form instead of slope-intercept form. Distribute to get $y - 3 = 4x - 28$ and add 3 to both sides to get $y = 4x - 25$.

7.3 Problems

(1) $\boxed{y = 6x - 21}$ Check: $y = 6(5) - 21 = 30 - 21 = 9$ agrees with $(5,9)$.

(2) $\boxed{y = -x + 4}$ Check: $(1)(-1) = -1$ and $b = 4$.

(3) $\boxed{y = -\frac{x}{3}}$ Check: $(3)\left(-\frac{1}{3}\right) = -1$ and $y = -\frac{0}{3} = 0$ agrees with $(0,0)$.

(4) $\boxed{y = -2x + 18}$ Check: $y = -2(9) + 18 = -18 + 18 = 0$ agrees with $(9,0)$.

Note: Exercises 4-5 give you the x-intercept (NOT the y-intercept).

(5) $\boxed{y = 9x - 54}$ Check: $y = 9(6) - 54 = 54 - 54 = 0$ agrees with $(6,0)$.

(6) $\boxed{y = \frac{4x}{3} - 40}$ Check: $\left(-\frac{3}{4}\right)\left(\frac{4}{3}\right) = -1$ and $y = \frac{4(12)}{3} - 40 = \frac{48}{3} - 40 = 16 - 40 = -24$ agrees with $(12,-24)$.

(7) $\boxed{y = -5x + 5}$ Check: $\left(\frac{1}{5}\right)(-5) = -1$ and $y = -5(-3) + 5 = 15 + 5 = 20$ agrees with $(-3,20)$.

(8) $\boxed{x = -3}$ Check: $y = 5$ is horizontal and $x = -3$ is vertical, so these lines are perpendicular, and $x = -3$ passes through every point where the x-coordinate is -3. Note: $y = 5$ has zero slope ($y = 0x + 5 = 5$) while $x = -3$ is infinitely steep.

(9) $\boxed{y = -\frac{5x}{2} + 14}$ Check: $\left(-\frac{5}{2}\right)\left(\frac{2}{5}\right) = -1$ and $y = -\frac{5(2)}{2} + 14 = -5 + 14 = 9$ agrees with $(2,9)$.

(10) $\boxed{y = 8x + 52}$ Check: $\left(-\frac{1}{8}\right)(8) = -1$ and $y = 8(-6) + 52 = -48 + 52 = 4$ agrees with $(-6, 4)$.

(11) $\boxed{y = -6x - 12}$ Check: $b = -12$. Note: $y = 9 - 6x$ is equivalent to $y = -6x + 9$.

(12) $\boxed{y = \frac{x}{7} - 2}$ Check: $(-7)\left(\frac{1}{7}\right) = -1$ and $y = \frac{14}{7} - 2 = 2 - 2 = 0$ agrees with $(14, 0)$.

Note: Exercise 12 gives you the x-intercept, whereas Exercise 11 gives you the y-intercept.

(13) $\boxed{y = -\frac{5x}{3} + \frac{4}{3}}$ Check: $\left(\frac{3}{5}\right)\left(-\frac{5}{3}\right) = -1$ and $y = -\frac{5}{3}\left(-\frac{1}{5}\right) + \frac{4}{3} = \frac{1}{3} + \frac{4}{3} = \frac{5}{3}$ agrees with $\left(-\frac{1}{5}, \frac{5}{3}\right)$.

(14) $\boxed{y = -\frac{x}{6} + \frac{1}{9}}$ Check: $y = -\frac{1}{6}\left(\frac{2}{3}\right) + \frac{1}{9} = -\frac{2}{18} + \frac{1}{9} = -\frac{1}{9} + \frac{1}{9} = 0$ agrees with $\left(\frac{2}{3}, 0\right)$.

Note: Exercises 14-15 give you the x-intercept (NOT the y-intercept).

(15) $\boxed{y = -\frac{x}{4} - 1}$ Check: $(4)\left(-\frac{1}{4}\right) = -1$ and $y = -\frac{(-4)}{4} - 1 = 1 - 1 = 0$ agrees with $(-4, 0)$.

(16) $\boxed{y = -2.5x + 0.8}$ Alternate answer: $\boxed{y = -\frac{5x}{2} + \frac{4}{5}}$. Check: $(0.4)(-2.5) = -1$, which may alternatively be expressed as $\left(\frac{2}{5}\right)\left(-\frac{5}{2}\right) = -1$, and $y = -2.5(0.8) + 0.8 = -2 + 0.8 = -1.2$ agrees with $(0.8, -1.2)$, which may alternatively be expressed as $= -\frac{5}{2}\left(\frac{4}{5}\right) + \frac{4}{5} = -2 + \frac{4}{5} = -\frac{10}{5} + \frac{4}{5} = -\frac{6}{5}$ and $\left(\frac{4}{5}, -\frac{6}{5}\right)$.

(17) $\boxed{x = -1}$ Check: $x = -3$ and $x = -1$ are both vertical lines (so they are parallel) and $x = -1$ passes through every point where the x-coordinate is -1.

(18) $\boxed{y = -\frac{x\sqrt{3}}{3} + 1}$ Alternate answer: $\boxed{y = -\frac{x}{\sqrt{3}} + 1}$ (but the first answer is in standard form with a rational denominator; recall Sec. 2.4). Check: $\left(\sqrt{3}\right)\left(-\frac{\sqrt{3}}{3}\right) = -1$ and $y = -\frac{\sqrt{3}\sqrt{3}}{3} + 1 = -\frac{3}{3} + 1 = -1 + 1 = 0$ agrees with $\left(\sqrt{3}, 0\right)$.

7.4 Problems

(1) $\boxed{900 \text{ feet}}$ from the station at 30 seconds; $\boxed{150 \text{ feet}}$ from the station at zero seconds; $\boxed{25 \text{ ft./s}}$; $\boxed{d = 25t + 150}$. Check: $d(12) = 25(12) + 150 = 300 + 150 = 450$, $d(20) = 25(20) + 150 = 500 + 150 = 650$, $d(30) = 25(30) + 150 = 750 + 150 =$

Answer Key

900, and $d(0) = 25(0) + 150 = 150$. Note: In this problem, the speed of the train is equal to the slope (since the slope is the change in distance over the change in time).

(2) $\boxed{\$150}$ for 60 cans; $\boxed{10 \text{ cans}}$ to break even; $\boxed{-\$6}$ is the loss for 8 cans; $\boxed{P = 3c - 30}$. Check: $P(25) = 3(25) - 30 = 75 - 30 = \$45$, $P(40) = 3(40) - 30 = 120 - 30 = \$90$, $P(10) = 3(10) - 30 = 30 - 30 = 0$, and $P(8) = 3(8) - 30 = 24 - 30 = -\$6$.

(3) $\boxed{1440°}$ for a decagon; $\boxed{180°}$ for a triangle; $\boxed{S = 180°N - 360°}$ which is equivalent to $S = (N - 2)180°$ (which is a well-known formula in geometry). Check: $S(5) = 180°(5) - 360° = 900° - 360° = 540°$, $S(7) = 180°(7) - 360° = 1260° - 360° = 900°$, $S(10) = 180°(10) - 360° = 1800° - 360° = 1440°$, and $S(3) = 180°(3) - 360° = 540° - 360° = 180°$.

(4) $\boxed{1080 \text{ ounces}}$ with 200 balls; $\boxed{480 \text{ ounces}}$ when empty; $\boxed{3 \text{ ounces}}$ is the weight of one ball; $\boxed{340 \text{ balls}}$ is the maximum capacity; $\boxed{W = 3N + 480}$. Check: $W(40) = 3(40) + 480 = 120 + 480 = 600$, $W(90) = 3(90) + 480 = 270 + 480 = 750$, $m = 3$ ounces per ball is the slope, and $W(340) = 3(340) + 480 = 1020 + 480 = 1500$.

(5) $\boxed{37 \text{ gallons}}$ at 32 minutes; $\boxed{53 \text{ gallons}}$ initially (at zero minutes); $\boxed{106 \text{ minutes}}$ is the time when the tub will be empty; $\boxed{V = -\frac{t}{2} + 53}$ (where V is for volume and where t is the time in minutes). Check: $V(10) = -\frac{10}{2} + 53 = -5 + 53 = 48$, $V(16) = -\frac{16}{2} + 53 = -8 + 53 = 45$, $V(32) = -\frac{32}{2} + 53 = -16 + 53 = 37$, $V(0) = -\frac{0}{2} + 53 = 53$, and $V(106) = -\frac{106}{2} + 53 = -53 + 53 = 0$.

(6) $\boxed{75 \text{ cents}}$ if an item sells for \$1.80; $\boxed{\$1.30}$ is the selling price if the profit is 50 cents; $\boxed{P = \frac{C}{2} - 15}$. Check: $P(60) = \frac{60}{2} - 15 = 30 - 15 = 15$, $P(90) = \frac{90}{2} - 15 = 45 - 15 = 30$, $P(180) = \frac{180}{2} - 15 = 90 - 15 = 75$, and $P(130) = \frac{130}{2} - 15 = 65 - 15 = 50$.

(7) $\boxed{10 \text{ m/s}}$ at 18 seconds; $\boxed{21 \text{ seconds}}$ is the time when the speed is zero; $\boxed{70 \text{ m/s}}$ initially (at zero seconds); $\boxed{v = -\frac{10}{3}t + 70}$. Check: $v(9) = -\frac{10}{3}(9) + 70 = -30 + 70 = 40$, $v(12) = -\frac{10}{3}(12) + 70 = -40 + 70 = 30$, $v(18) = -\frac{10}{3}(18) + 70 = -60 + 70 = 10$, $v(21) = -\frac{10}{3}(21) + 70 = -70 + 70 = 0$, and $v(0) = -\frac{10}{3}(0) + 70 = 70$.

(8) $\boxed{\frac{10}{3} \text{ eV}}$ at 800 THz; $\boxed{1200 \text{ THz}}$ is the frequency for 6 eV; $\boxed{K(f) = \frac{f}{150} - 2}$ where the

kinetic energy is in eV and the frequency is in THz. Check: $K(0) = \frac{0}{150} - 2 = -2, K(300) =$

$\frac{300}{150} - 2 = 2 - 2 = 0, K(800) = \frac{800}{150} - 2 = \frac{16}{3} - \frac{6}{3} = \frac{10}{3}$, and $K(1200) = \frac{1200}{150} - 2 = 8 - 2 = 6$.

(9) $\boxed{17 \text{ seconds}}$ is the time when the rocket runs out of fuel; $\boxed{850 \text{ kg}}$ is the initial fuel (at

zero seconds); $\boxed{m(t) = -50t + 850}$. Check: $m(7) = -50(7) + 850 = -350 + 850 =$

$500, m(12) = -50(12) + 850 = -600 + 850 = 250, m(17) = -50(17) + 850 =$

$-850 + 850 = 0$, and $m(0) = -50(0) + 850 = 850$.

(10) $\boxed{\text{day } 48}$ is when the stocks will both have a value of \$107; $\boxed{A = 2t + 11}$,

$\boxed{B = -t + 155}$. Check: $A(22) = 2(22) + 11 = 44 + 11 = 55, A(28) = 2(28) + 11 =$

$56 + 11 = 67, B(14) = -14 + 155 = 141, B(30) = -30 + 155 = 125, A(48) =$

$2(48) + 11 = 96 + 11 = 107$, and $B(48) = -48 + 155 = 107$.

7.5 Problems

(1) $\boxed{(3,8)}, R = \boxed{7}$ Check: $R^2 = 7^2 = 49$.

(2) $\boxed{(0,0)}, R = \boxed{\sqrt{5}}$ Check: $R^2 = \left(\sqrt{5}\right)^2 = 5$.

(3) $\boxed{(-9,4)}, R = \boxed{2\sqrt{2}}$ Check: $R^2 = \left(2\sqrt{2}\right)^2 = 2^2\left(\sqrt{2}\right)^2 = 4(2) = 8$.

Notes: $\sqrt{8} = \sqrt{4}\sqrt{2} = 2\sqrt{2}$ and $a = -9$ because $x - a = x - (-9) = x + 9$.

(4) $x^2 + 8x + y^2 = 9 \boxed{(-4,0)}, R = \boxed{5}$ Check: $R^2 = 5^2 = 25$. See below.

Notes: Add 16 to both sides of the given equation: $x^2 + 8x + 16 + y^2 = 25$. Factor

the left-hand side: $(x + 4)^2 + y^2 = 25$. $a = -4$ because $x - a = x - (-4) = x + 4$.

R isn't 3 because $x^2 + 8x + y^2 = 9$ isn't in standard form. See Example 2.

(5) $x^2 - 16x + y^2 + 10y = 32 \boxed{(8,-5)}, R = \boxed{11}$ Check: $R^2 = 11^2 = 121$. See below.

Notes: Add 64 and add 25 to both sides of the given equation: $x^2 - 16 + 64 + y^2 + 10y + 25$

$= 32 + 64 + 25$. Factor and simplify: $(x - 8)^2 + (y + 5)^2 = 121$.

(6) $2x^2 + 2y^2 = 54 + 24y - 36x \boxed{(-9,6)}, R = \boxed{12}$ Check: $R^2 = 12^2 = 144$. See below.

Notes: Divide by 2 and rearrange terms: $x^2 + 18x + y^2 - 12y = 27$. Add 81 and add 36

to both sides: $x^2 + 18x + 81 + y^2 - 12y + 36 = 27 + 81 + 36$. Factor and simplify:

$(x + 9)^2 + (y - 6)^2 = 144$.

(7) $\boxed{(1,6)}$ and $\boxed{\left(-\frac{21}{5}, -\frac{22}{5}\right)}$ Check: $x^2 + y^2 = 1^2 + 6^2 = 1 + 36 = 37, y = 2(1) + 4 =$

$2 + 4 = 6, x^2 + y^2 = \left(-\frac{21}{5}\right)^2 + \left(-\frac{22}{5}\right)^2 = \frac{441}{25} + \frac{484}{25} = \frac{925}{25} = 37$, and $y = 2\left(-\frac{21}{5}\right) + 4 =$

$-\frac{42}{5} + \frac{20}{5} = -\frac{22}{5}$.

(8) $\boxed{\left(\frac{5}{4}, -\frac{3\sqrt{7}}{4}\right)}$ and $\boxed{\left(\frac{5}{4}, \frac{3\sqrt{7}}{4}\right)}$ Check: $(x-1)^2 + y^2 = \left(\frac{5}{4}-1\right)^2 + \left(\pm\frac{3\sqrt{7}}{4}\right)^2 = \left(\frac{1}{4}\right)^2 + \frac{3^2(\sqrt{7})^2}{4^2} =$

$\frac{1}{16} + \frac{9(7)}{16} = \frac{1}{16} + \frac{63}{16} = \frac{64}{16} = 4$ and $(x+1)^2 + y^2 = \left(\frac{5}{4}+1\right)^2 + \left(\pm\frac{3\sqrt{7}}{4}\right)^2 = \left(\frac{9}{4}\right)^2 + \frac{3^2(\sqrt{7})^2}{4^2} =$

$\frac{81}{16} + \frac{9(7)}{16} = \frac{81}{16} + \frac{63}{16} = \frac{144}{16} = 9$. Note: $\frac{\sqrt{63}}{4} = \frac{\sqrt{9}\sqrt{7}}{4} = \frac{3\sqrt{7}}{4}$.

(9) $\boxed{(-3,-3)}$ and $\boxed{(6,6)}$ Check: $x^2 + (y-3)^2 = (-3)^2 + (-3-3)^2 = 9 + (-6)^2 =$

$9 + 36 = 45, y = x = -3, x^2 + (y-3)^2 = 6^2 + (6-3)^2 = 36 + 3^2 = 36 + 9 = 45,$

and $y = x = 6$.

(10) $\boxed{(-4,-5)}$ and $\boxed{(-3,-6)}$ Check: $x^2 + (y+2)^2 = (-4)^2 + (-5+2)^2 = 16 + (-3)^2 =$

$16 + 9 = 25, y = -x - 9 = -(-4) - 9 = 4 - 9 = -5, x^2 + (y+2)^2 = (-3)^2 + (-6+2)^2 =$

$9 + (-4)^2 = 9 + 16 = 25$, and $y = -x - 9 = -(-3) - 9 = 3 - 9 = -6$.

7.6 Problems

(1) $\boxed{(\pm 3, 0)}$ Notes: $a = 5, b = 4, a > b, c = \sqrt{a^2 - b^2} = \sqrt{5^2 - 4^2} = \sqrt{25 - 16} = \sqrt{9} = 3$.

(2) $\boxed{(0, \pm 1)}$ Notes: $a = \sqrt{3}, b = 2, a < b, c = \sqrt{b^2 - a^2} = \sqrt{2^2 - (\sqrt{3})^2} = \sqrt{4 - 3} = \sqrt{1} = 1$.

(3) $\boxed{(\pm 2\sqrt{2}, 0)}$ Notes: $a = 3, b = 1, a > b, c = \sqrt{a^2 - b^2} = \sqrt{3^2 - 1^2} = \sqrt{9 - 1} = \sqrt{8} =$

$\sqrt{4}\sqrt{2} = 2\sqrt{2}$.

(4) $\boxed{(\pm 2, 0)}$ Notes: $a = 2\sqrt{2}, b = 2, a > b, c = \sqrt{a^2 - b^2} = \sqrt{(2\sqrt{2})^2 - 2^2} = \sqrt{2^2(\sqrt{2})^2 - 4} =$

$\sqrt{4(2) - 4} = \sqrt{8 - 4} = \sqrt{4} = 2$. Divide $10x^2 + 6y^2 = 60$ by 60 to get $\frac{x^2}{6} + \frac{y^2}{10} = 1$.

(5) $\boxed{(0, \pm 2)}$ Notes: $a = \sqrt{6}, b = \sqrt{10}, a < b, c = \sqrt{b^2 - a^2} = \sqrt{(\sqrt{10})^2 - (\sqrt{6})^2} = \sqrt{10 - 6} =$

$\sqrt{4} = 2$.

(6) $\boxed{(\pm 5, 0)}$ Notes: $a = 13, b = 12, a > b, c = \sqrt{a^2 - b^2} = \sqrt{13^2 - 12^2} = \sqrt{169 - 144} =$

$\sqrt{25} = 5$.

(7) $\boxed{(0,\pm 8)}$ Notes: $a = 15$, $b = 17$, $a < b$, $c = \sqrt{b^2 - a^2} = \sqrt{17^2 - 15^2} = \sqrt{289 - 225} = \sqrt{64} = 8$.

(8) $\boxed{(-1,2)}$ and $\boxed{(7,2)}$ Notes: $a = 5$, $b = 3$, $a > b$, $c = \sqrt{a^2 - b^2} = \sqrt{5^2 - 3^2} = \sqrt{25 - 9} = \sqrt{16} = 4$. This shifted ellipse is centered at $(3,2)$; you can deduce this by reviewing the equations for circles in Sec. 7.5. The distance from the center of the ellipse to each focus is $c = 4$, such that the foci lie at $(3 - 4, 2)$ and $(3 + 4, 2)$, which simplify to $(-1,2)$ and $(7,2)$.

7.7 Problems

(1) $\boxed{\left(2, -\frac{11}{12}\right)}$, $x = 2$, and $y = -\frac{13}{12}$. Notes: $a = 3$, $b = 2$, $c = -1$, $-1 + \frac{1}{4(3)} = -\frac{12}{12} + \frac{1}{12} = -\frac{11}{12}$, and $-1 - \frac{1}{4(3)} = -\frac{12}{12} - \frac{1}{12} = -\frac{13}{12}$. Note the signs of b and c in $y = a(x - b)^2 + c$.

(2) $\boxed{(9, -6)}$, $y = -6$, and $x = 7$. Notes: $a = \frac{1}{4}$ (such that $4a = 1$), $b = -6$, and $c = 8$. The roles of x and y are reversed in this problem (like Example 2).

(3) $\boxed{\left(0, -\frac{3}{4}\right)}$, $x = 0$, and $y = \frac{3}{4}$. Notes: $a = -\frac{1}{3}$ (such that $4a = -\frac{4}{3}$ and $\frac{1}{4a} = -\frac{3}{4}$), $b = 0$, and $c = 0$.

(4) $\boxed{\left(-\frac{5}{7}, 0\right)}$, $y = 0$, and $x = -\frac{11}{14}$. Notes: $a = 7$, $b = 0$, and $c = -\frac{3}{4}$. The roles of x and y are reversed in this problem (like Example 2). $-\frac{3}{4} + \frac{1}{4(7)} = -\frac{21}{28} + \frac{1}{28} = -\frac{20}{28} = -\frac{5}{7}$ and $-\frac{3}{4} - \frac{1}{4(7)} = -\frac{21}{28} - \frac{1}{28} = -\frac{22}{28} = -\frac{11}{14}$.

(5) $\boxed{\left(-2, -\frac{455}{24}\right)}$, $x = -2$, and $y = -\frac{457}{24}$. Notes: $a = 6$, $b = -2$, $c = -19$, $-19 + \frac{1}{4(6)} = -\frac{456}{24} + \frac{1}{24} = -\frac{455}{24}$, and $-19 - \frac{1}{4(6)} = -\frac{456}{24} - \frac{1}{24} = -\frac{457}{24}$. Check: $y = 6(x + 2)^2 - 19 = 6(x^2 + 4x + 4) - 19 = 6x^2 + 24x + 24 - 19 = 6x^2 + 24x + 5$.

(6) $\boxed{\left(3, -\frac{1079}{20}\right)}$, $x = 3$, and $y = -\frac{1081}{20}$. Notes: $a = 5$, $b = 3$, $c = -54$, $-54 + \frac{1}{4(5)} = -\frac{1080}{20} + \frac{1}{20} = -\frac{1079}{20}$, and $-54 - \frac{1}{4(5)} = -\frac{1080}{20} - \frac{1}{20} = -\frac{1081}{20}$. Check: $y = 5(x - 3)^2 - 54 = 5(x^2 - 6x + 9) - 54 = 5x^2 - 30x + 45 - 54 = 5x^2 - 30x - 9$.

(7) $\boxed{(-2,4)}$ and $\boxed{(3,9)}$ Check: $y = -2 + 6 = 4$, $y = x^2 = (-2)^2 = 4$, $y = 3 + 6 = 9$, and $y = x^2 = 3^2 = 9$.

(8) $(-4,-7)$ and $\left(\frac{3}{2},\frac{19}{2}\right)$ Check: $y = 3(-4) + 5 = -12 + 5 = -7, y = 2(-4)^2 + 8(-4) - 7$

$= 2(16) - 32 - 7 = 32 - 39 = -7, y = 3\left(\frac{3}{2}\right) + 5 = \frac{9}{2} + \frac{10}{2} = \frac{19}{2}$, and $y = 2\left(\frac{3}{2}\right)^2 + 8\left(\frac{3}{2}\right) - 7$

$= 2\left(\frac{9}{4}\right) + 12 - 7 = \frac{9}{2} + 5 = \frac{9}{2} + \frac{10}{2} = \frac{19}{2}$.

(9) $\left(-\frac{37}{4},-\frac{1}{2}\right)$ and $\left(-\frac{29}{3},-\frac{4}{3}\right)$ Check: $y = 2\left(-\frac{37}{4}\right) + 18 = -\frac{37}{2} + \frac{36}{2} = -\frac{1}{2}, x =$

$3\left(-\frac{1}{2}\right)^2 + 6\left(-\frac{1}{2}\right) - 7 = \frac{3}{4} - 3 - 7 = \frac{3}{4} - \frac{12}{4} - \frac{28}{4} = -\frac{37}{4}, y = 2\left(-\frac{29}{3}\right) + 18 = -\frac{58}{3} + \frac{54}{3} =$

$-\frac{4}{3}$, and $x = 3\left(-\frac{4}{3}\right)^2 + 6\left(-\frac{4}{3}\right) - 7 = \frac{16}{3} - \frac{24}{3} - \frac{21}{3} = -\frac{29}{3}$.

(10) $\left(\frac{4}{3},\frac{17}{9}\right)$ and $(5,8)$ Check: $y = 2\left(\frac{4}{3}\right)^2 - 11\left(\frac{4}{3}\right) + 13 = 2\left(\frac{16}{9}\right) - \frac{44}{3} + 13 = \frac{32}{9} -$

$\frac{132}{9} + \frac{117}{9} = \frac{17}{9}, y = -\left(\frac{4}{3}\right)^2 + 8\left(\frac{4}{3}\right) - 7 = -\frac{16}{9} + \frac{32}{3} - 7 = -\frac{16}{9} + \frac{96}{9} - \frac{63}{9} = \frac{17}{9}, y =$

$2(5)^2 - 11(5) + 13 = 50 - 42 = 8$, and $y = -(5)^2 + 8(5) - 7 = -25 + 40 - 7 = 8$.

(11) $(-2\sqrt{2},1)$ and $(2\sqrt{2},1)$ are the real solutions. Note: $\sqrt{8} = \sqrt{4}\sqrt{2} = 2\sqrt{2}$.

Check: $x^2 + y^2 = \left(\pm 2\sqrt{2}\right)^2 + 1^2 = (\pm 2)^2\left(\sqrt{2}\right)^2 + 1 = 4(2) + 1 = 8 + 1 = 9$, and

$y = \frac{\left(\pm 2\sqrt{2}\right)^2}{8} = \frac{(\pm 2)^2\left(\sqrt{2}\right)^2}{8} = \frac{4(2)}{8} = \frac{8}{8} = 1$.

(12) $(0,0)$ and $\left(\frac{1}{8},\frac{1}{16}\right)$ Check: $y = 4x^2 = 4(0)^2 = 0, x = 32y^2 = 32(0)^2 = 0, y =$

$4\left(\frac{1}{8}\right)^2 = \frac{4}{64} = \frac{1}{16}$, and $x = 32y^2 = 32\left(\frac{1}{16}\right)^2 = \frac{32}{256} = \frac{1}{8}$.

7.8 Problems

(1) $(\pm 10,0)$ and $y = \pm\frac{4x}{3}$. Notes: $a = 6, b = 8, c = \sqrt{6^2 + 8^2} = \sqrt{36 + 64} = \sqrt{100} =$

$10, \frac{8}{6}$ reduces to $\frac{4}{3}$.

(2) $(\pm 4,0)$ and $y = \pm\frac{x\sqrt{3}}{3}$. Notes: $a = \sqrt{12} = \sqrt{4}\sqrt{3} = 2\sqrt{3}, b = 2, c = \sqrt{\left(2\sqrt{3}\right)^2 + 2^2} =$

$\sqrt{2^2\left(\sqrt{3}\right)^2 + 4} = \sqrt{4(3) + 4} = \sqrt{12 + 4} = \sqrt{16} = 4, y = \pm\frac{x}{\sqrt{3}} = \pm\frac{x}{\sqrt{3}}\frac{\sqrt{3}}{\sqrt{3}} = \pm\frac{x\sqrt{3}}{3}$.

(3) $\boxed{(0,\pm3\sqrt{3})}$ and $\boxed{x=\pm y\sqrt{2}}$ or $\boxed{y=\pm\frac{x\sqrt{2}}{2}}$. Notes: $a=3, b=\sqrt{18}=\sqrt{9}\sqrt{2}=3\sqrt{2}$,

$c=\sqrt{3^2+\left(3\sqrt{2}\right)^2}=\sqrt{9+3^2\left(\sqrt{2}\right)^2}=\sqrt{9+9(2)}=\sqrt{9+18}=\sqrt{27}=\sqrt{9}\sqrt{3}=3\sqrt{3}$. In

this problem, the roles of x and y are reversed (like Example 2). We rationalized the

denominator in $y=\pm\frac{x}{\sqrt{2}}=\pm\frac{x}{\sqrt{2}}\frac{\sqrt{2}}{\sqrt{2}}=\pm\frac{x\sqrt{2}}{2}$.

(4) $\boxed{(\pm2,0)}$ and $\boxed{y=\pm\frac{x\sqrt{3}}{3}}$. Notes: $a=\sqrt{3}, b=1, c=\sqrt{\left(\sqrt{3}\right)^2+1^2}=\sqrt{3+1}=\sqrt{4}=2$,

$y=\pm\frac{x}{\sqrt{3}}=\pm\frac{x}{\sqrt{3}}\frac{\sqrt{3}}{\sqrt{3}}=\pm\frac{x\sqrt{3}}{3}$.

(5) $\boxed{(\pm13,0)}$ and $\boxed{y=\pm\frac{5x}{12}}$. Notes: $a=12, b=5, c=\sqrt{12^2+5^2}=\sqrt{144+25}=$

$\sqrt{169}=13$.

(6) $\boxed{(0,\pm17)}$ and $\boxed{x=\pm\frac{15y}{8}}$ or $\boxed{y=\pm\frac{8x}{15}}$. Notes: $a=8, b=15, c=\sqrt{8^2+15^2}=$

$\sqrt{64+225}=\sqrt{289}=17$. In this problem, the roles of x and y are reversed (like

Example 2).

(7) $4x^2-3y^2=12$ $\boxed{(\pm\sqrt{7},0)}$ and $\boxed{y=\pm\frac{2x\sqrt{3}}{3}}$. Notes: $a=\sqrt{3}, b=2, c=\sqrt{\left(\sqrt{3}\right)^2+2^2}=$

$\sqrt{3+4}=\sqrt{7}$. Divide $4x^2-3y^2=12$ by 12 to get $\frac{x^2}{3}-\frac{y^2}{4}=1$. We rationalized the

denominator in $y=\pm\frac{2x}{\sqrt{3}}=\pm\frac{2x}{\sqrt{3}}\frac{\sqrt{3}}{\sqrt{3}}=\pm\frac{2x\sqrt{3}}{3}$. Note that $a=\sqrt{3}$ and $b=2$ because in standard

form the equation is $\frac{x^2}{3}-\frac{y^2}{4}=1$. If you make the mistake of trying to identify a and b from

the given equation (instead of standard form), you can easily get these backwards.

(8) $\boxed{(-\sqrt{2},\sqrt{2})}$, $\boxed{(\sqrt{2},-\sqrt{2})}$, $\boxed{x=0}$, and $\boxed{y=0}$. Notes: $a=\sqrt{2}, b=\sqrt{2}, c=\sqrt{\left(\sqrt{2}\right)^2+\left(\sqrt{2}\right)^2}$

$=\sqrt{2+2}=\sqrt{4}=2$. The distance from the origin to either focus is $\sqrt{\left(\sqrt{2}\right)^2+\left(\sqrt{2}\right)^2}=$

$\sqrt{2+2}=\sqrt{4}=2$, which agrees with c. Note that $y=-\frac{1}{x}$ has branches in Quadrants II and

IV, whereas $y=\frac{1}{x}$ (discussed in the chapter) has branches in Quadrants I and III. This

problem features a **rotated rectangular hyperbola** (discussed in the first paragraph of the

chapter).

7.9 Problems

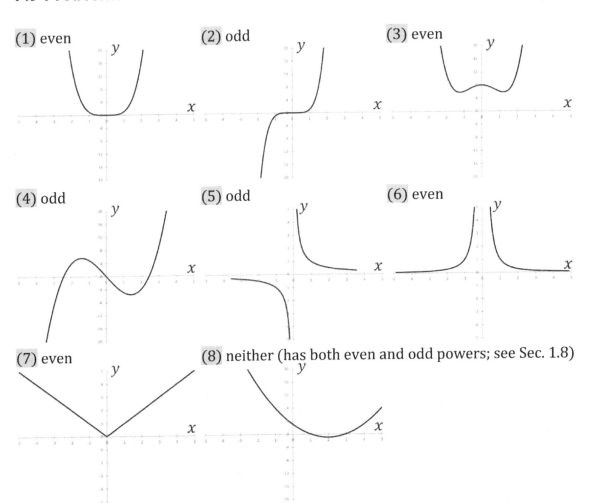

(1) even

(2) odd

(3) even

(4) odd

(5) odd

(6) even

(7) even

(8) neither (has both even and odd powers; see Sec. 1.8)

8 Inequalities

8.1 Problems

(1) $\boxed{x > 11}$ Check the long way: $-x < -11 \to 0 < x - 11 \to 11 < x$.

Note: We are checking the answers the long way, but you are expected to solve the problems the short way by **dividing both sides by a negative number** at some point in your solution.

(2) $\boxed{-5 < x}$ Check the long way: $5 > -x \to 5 + x > 0 \to x > -5$.

(3) $\boxed{x < -7}$ Check the long way: $-3x > 21 \to 0 > 3x + 21 \to -21 > 3x \to -7 > x$.

(4) $\boxed{x > 2}$ Check the long way: $-6x < -12 \to 0 < 6x - 12 \to 12 < 6x \to 2 < x$.

(5) $\boxed{x \geq -6}$ Check the long way: $5x \leq 9x + 24 \to 0 \leq 4x + 24 \to -24 \leq 4x \to -6 \leq x$.

(6) $\boxed{8 > x}$ Check the long way: $3x - 56 < -4x \to 7x - 56 < 0 \to 7x < 56 \to x < 8$.

(7) $\boxed{4 < x}$ Check the long way: $-4x - 20 > -9x \to 5x - 20 > 0 \to 5x > 20 \to x > 4$.

(8) $\boxed{-5 \leq x}$ Check the long way: $35 + 8x \geq x \to 35 + 7x \geq 0 \to 7x \geq -35 \to x \geq -5$.

(9) $\boxed{x > \frac{1}{2}}$ Check the long way: $-x < -1 + x \to 0 < -1 + 2x \to 1 < 2x \to \frac{1}{2} < x$.

(10) $\boxed{10 > x}$ Check the long way: $8x - 100 < -2x \to 10x - 100 < 0 \to 10x < 100 \to x < 10$.

(11) $\boxed{x < 3}$ Check the long way: $-3(x - 5) > 6 \to -3x + 15 > 6 \to -3x + 9 > 0 \to 9 > 3x \to 3 > x$.

(12) $\boxed{x > 6}$ Check the long way: $-6(x - 5) < -2(9 - x) \to -6x + 30 < -18 + 2x \to 48 < 8x \to 6 < x$. Tip: In the first step, divide both sides by -2.

(13) $\boxed{-5 \geq x}$ Check the long way: $9 + 5x + 6 \leq 2x \to 15 + 5x \leq 2x \to 15 + 3x \leq 0 \to 3x \leq -15 \to x \leq -5$.

(14) $\boxed{x < 7}$ Check the long way: $-3x - 8x > -9x - 14 \to -11x > -9x - 14 \to 0 > 2x - 14 \to 14 > 2x \to 7 > x$.

8.2 Problems

(1) $\boxed{x < 7}$ or $\boxed{x > -7}$; equivalent to $\boxed{|x| < 7}$. Note: $(\pm 7)^2 = 49$.

(2) $\boxed{x \geq \sqrt{3}}$ or $\boxed{x \leq -\sqrt{3}}$; equivalent to $\boxed{|x| \geq \sqrt{3}}$. Note: $6(\pm\sqrt{3})^2 = 6(3) = 18$.

(3) $\boxed{6 < x}$ or $\boxed{-6 > x}$; equivalent to $\boxed{|x| > 6}$. Note: $(\pm 6)^2 - 16 = 36 - 16 = 20$.

(4) $\boxed{x \in \mathbb{R}}$. Note: Any real value of x satisfies $x^2 > -2$.

(5) $\boxed{x > 2}$ or $\boxed{x < -2}$; equivalent to $\boxed{|x| > 2}$. Notes: $7(\pm 2)^2 = 7(4) = 28$ and $2(\pm 2)^2 + 20 = 2(4) + 20 = 8 + 20 = 28$.

(6) $\boxed{x < 4}$ or $\boxed{x > -4}$; equivalent to $\boxed{|x| < 4}$. Note: $4(\pm 4)^2 - 21 = 4(16) - 21 = 64 - 21 = 43$.

(7) $\boxed{x < 5}$ or $\boxed{x > -5}$; equivalent to $\boxed{|x| < 5}$. Notes: $-8(\pm 5)^2 = -8(25) = -200$. $-8x^2 > -200$ becomes $8x^2 < 200$ (recall Sec. 8.1).

(8) $\boxed{2\sqrt{3} \le x}$ or $\boxed{-2\sqrt{3} \ge x}$; equivalent to $\boxed{|x| \ge 2\sqrt{3}}$. Notes: $24 - \left(2\sqrt{3}\right)^2 = 24 - 2^2\left(\sqrt{3}\right)^2 = 24 - 4(3) = 24 - 12$, $\left(2\sqrt{3}\right)^2 = 2^2\left(\sqrt{3}\right)^2 = 4(3) = 12$, and $\sqrt{12} = \sqrt{4}\sqrt{3} = 2\sqrt{3}$.

(9) $\boxed{x \in \mathbb{R}}$. Note: Any real value of x satisfies $-x^2 < \frac{4}{7}$, which is equivalent to $x^2 > -\frac{4}{7}$.

(10) $\boxed{x < \frac{1}{2}}$ or $\boxed{x > -\frac{1}{2}}$; equivalent to $\boxed{|x| < \frac{1}{2}}$. Notes: $-4\left(\frac{1}{2}\right)^2 = -4\left(\frac{1}{4}\right) = -1$. $-4x^2 > -1$ becomes $4x^2 < 1$ (recall Sec. 8.1).

(11) $\boxed{x \ge 3}$ or $\boxed{x \le -3}$; equivalent to $\boxed{|x| \ge 3}$. Note: $9(\pm 3)^2 - 12 = 9(9) - 12 = 81 - 12 = 69$ and $24 + 5(\pm 3)^2 = 24 + 5(9) = 24 + 45 = 69$.

(12) $\boxed{\sqrt{5} < x}$ or $\boxed{-\sqrt{5} > x}$; equivalent to $\boxed{|x| > \sqrt{5}}$. Note: $7 - 2\left(\pm\sqrt{5}\right)^2 = 7 - 2(5) = 7 - 10 = -3$ and $2 - \left(\pm\sqrt{5}\right)^2 = 2 - 5 = -3$.

(13) $\boxed{3\sqrt{3} < x}$ or $\boxed{-3\sqrt{3} > x}$; equivalent to $\boxed{|x| > 3\sqrt{3}}$. Note: $-3\left(\pm 3\sqrt{3}\right)^2 - 83 = -3(3)^2\left(\sqrt{3}\right)^2 - 83 = -3(9)(3) - 83 = -81 - 83 = -164$, $-6\left(\pm 3\sqrt{3}\right)^2 - 2 = -6(3)^2\left(\sqrt{3}\right)^2 - 2 = -6(9)(3) - 2 = -162 - 2 = -164$, and $\sqrt{27} = \sqrt{9}\sqrt{3} = 3\sqrt{3}$. Tip: Multiply $-3x^2 - 83 > -6x^2 - 2$ by -1 to get $3x^2 + 83 < 6x^2 + 2$ (Sec. 8.1).

(14) $\boxed{x \le \sqrt{10}}$ or $\boxed{x \ge -\sqrt{10}}$; equivalent to $\boxed{|x| \le \sqrt{10}}$. Note: $4\left(\pm\sqrt{10}\right)^2 - 29 = 4(10) - 29 = 40 - 29 = 11$ and $-6\left(\pm\sqrt{10}\right)^2 + 71 = -6(10) + 71 = -60 + 71 = 11$.

8.3 Problems

(1) $\boxed{-2 < x < 9}$; equivalent to $\boxed{x \in (-2, 9)}$. Note: $(x + 2)(x - 9) = x^2 - 9x + 2x - 18 = x^2 - 7x - 18$.

(2) $\boxed{x < 5}$ or $\boxed{x > 8}$; equivalent to $\boxed{x \in (-\infty, 5) \cup (8, \infty)}$. Note: $(x - 5)(x - 8) = x^2 - 8x - 5x + 40 = x^2 - 13x + 40$.

(3) $\boxed{x < -\frac{4}{3}}$ or $\boxed{x > \frac{1}{2}}$; equivalent to $\boxed{x \in \left(-\infty, -\frac{4}{3}\right) \cup \left(\frac{1}{2}, \infty\right)}$. Note: $(3x + 4)(2x - 1) = 6x^2 - 3x + 8x - 4 = 6x^2 + 5x - 4$.

(4) $\boxed{-\frac{5}{3} < x < -\frac{3}{5}}$; equivalent to $\boxed{x \in \left(-\frac{5}{3}, -\frac{3}{5}\right)}$. Note: $(3x + 5)(5x + 3) = 15x^2 + 25x + 9x + 15 = 15x^2 + 34x + 15$.

(5) $\boxed{x < \frac{3}{8}}$ or $\boxed{x > \frac{5}{2}}$; equivalent to $\boxed{x \in \left(-\infty, \frac{3}{8}\right) \cup \left(\frac{5}{2}, \infty\right)}$. Note: $(8x - 3)(2x - 5) = 16x^2 - 40x - 6x + 15 = 16x^2 - 46x + 15$.

(6) $\boxed{-\frac{4}{5} < x < \frac{5}{4}}$; equivalent to $\boxed{x \in \left(-\frac{4}{5}, \frac{5}{4}\right)}$. Note: $(5x + 4)(4x - 5) = 20x^2 + 16x - 25x - 20 = 20x^2 - 9x - 20$.

(7) $\boxed{-5 < x < -\frac{9}{8}}$; equivalent to $\boxed{x \in \left(-5, -\frac{9}{8}\right)}$. Note: $(8x + 9)(x + 5) = 8x^2 + 40x + 9x + 45 = 8x^2 + 49x + 45$.

(8) $\boxed{x < -\frac{3}{4}}$ or $\boxed{x > \frac{5}{6}}$; equivalent to $\boxed{x \in \left(-\infty, -\frac{3}{4}\right) \cup \left(\frac{5}{6}, \infty\right)}$. Note: $(4x + 3)(6x - 5) = 24x^2 + 18x - 20x - 15 = 24x^2 - 2x - 15$.

(9) $\boxed{x < \frac{7}{3}}$ or $\boxed{x > \frac{7}{3}}$; a simpler answer is $\boxed{x \neq \frac{7}{3}}$. Note: $(3x - 7)(3x - 7) = 9x^2 - 21x - 21x + 49 = 9x^2 - 42x + 49$.

(10) $\boxed{-2 < x < \frac{3}{5}}$; equivalent to $\boxed{x \in \left(-2, \frac{3}{5}\right)}$. Note: $(6x + 12)(5x - 3) = 30x^2 - 18x + 60x - 36 = 30x^2 + 42x - 36$.

8.4 Problems

(1) $\boxed{0 < x < 6}$; equivalent to $\boxed{x \in (0,6)}$. Note: $\frac{1}{x} > \frac{1}{6}$ is like Example 1.

(2) $\boxed{x < 0}$ or $\boxed{x > 2}$; equivalent to $\boxed{x \in (-\infty, 0) \cup (2, \infty)}$. Note: $\frac{1}{x} < \frac{1}{2}$ is like Example 2.

(3) $\boxed{-4 < x < 0}$; equivalent to $\boxed{x \in (-4,0)}$. Note: $\frac{1}{x} < -\frac{1}{4}$ is like Example 3.

(4) $\boxed{x < -3}$ or $\boxed{x > 0}$; equivalent to $\boxed{x \in (-\infty, -3) \cup (0, \infty)}$. Note: $\frac{1}{x} > -\frac{1}{3}$ is like Example 4.

(5) $\boxed{x < 0}$ or $\boxed{x > 8}$; equivalent to $\boxed{x \in (-\infty, 0) \cup (8, \infty)}$. Note: $\frac{1}{8} > \frac{1}{x}$ is equivalent to $\frac{1}{x} < \frac{1}{8}$, which is like Example 2.

(6) $\boxed{0 < x < 10}$; equivalent to $\boxed{x \in (0, 10)}$. Note: $\frac{1}{10} < \frac{1}{x}$ is equivalent to $\frac{1}{x} > \frac{1}{10}$, which is like Example 1.

(7) $\boxed{x < -12}$ or $\boxed{x > 0}$; equivalent to $\boxed{x \in (-\infty, -12) \cup (0, \infty)}$. Note: $-\frac{1}{12} < \frac{1}{x}$ is equivalent to $\frac{1}{x} > -\frac{1}{12}$, which is like Example 4.

(8) $\boxed{-15 < x < 0}$; equivalent to $\boxed{x \in (-15,0)}$. Note: $-\frac{1}{15} > \frac{1}{x}$ is equivalent to $\frac{1}{x} < -\frac{1}{15}$, which is like Example 3.

(9) $\boxed{0 < x < \frac{1}{3}}$; equivalent to $\boxed{x \in \left(0, \frac{1}{3}\right)}$. Notes: $\frac{1}{x} > 3$ is like Example 1. When we take the reciprocal of both sides, the reciprocal of 3 equals $\frac{1}{3}$.

(10) $\boxed{x < -5}$ or $\boxed{x > 0}$; equivalent to $\boxed{x \in (-\infty, -5) \cup (0, \infty)}$. Notes: First multiply by -1 on both sides, which reverses the direction of the inequality (Sec. 8.1): $\frac{1}{x} > -\frac{1}{5}$. This is like Example 4.

8.5 Problems

(1) $\boxed{0 < x < 15}$; equivalent to $\boxed{x \in (0, 15)}$. Note: $\frac{18}{15} = \frac{24}{20} = \frac{6}{5}$.

(2) $\boxed{x < 0}$ or $\boxed{x > 56}$; equivalent to $\boxed{x \in (-\infty, 0) \cup (56, \infty)}$. Note: $\frac{16}{56} = \frac{10}{35} = \frac{2}{7}$.

(3) $\boxed{x < -14}$ or $\boxed{x > 0}$; equivalent to $\boxed{x \in (-\infty, -14) \cup (0, \infty)}$. Notes: $-\frac{6}{14} = -\frac{3}{7}$. For $x < -14$, $\frac{6}{x}$ is less negative than $-\frac{3}{7}$, meaning that $\frac{6}{x}$ is closer to zero.

(4) $\boxed{-9 < x < 0}$; equivalent to $\boxed{x \in (-9,0)}$. Notes: $-\frac{12}{9} = -\frac{72}{54} = -\frac{4}{3}$.

For $-9 < x < 0$, $\frac{12}{x}$ is more negative than $-\frac{72}{54}$, meaning that $\frac{12}{x}$ is farther from zero.

(5) $\boxed{x < 0}$ or $\boxed{x > 64}$; equivalent to $\boxed{x \in (-\infty, 0) \cup (64, \infty)}$. Note: $\frac{3}{8} = \frac{24}{64}$.

(6) $\boxed{0 < x < 20}$; equivalent to $\boxed{x \in (0, 20)}$. Note: $\frac{27}{15} = \frac{36}{20} = \frac{9}{5}$.

(7) $\boxed{x < -28}$ or $\boxed{x > 0}$; equivalent to $\boxed{x \in (-\infty, -28) \cup (0, \infty)}$. Notes: $-\frac{39}{42} = -\frac{26}{28} = -\frac{13}{14}$.

For $x < -28$, $\frac{26}{x}$ is less negative than $-\frac{39}{42}$, meaning that $\frac{26}{x}$ is closer to zero.

(8) $\boxed{-99 < x < 0}$; equivalent to $\boxed{x \in (-99, 0)}$. Notes: $-\frac{11}{9} = -\frac{121}{99}$.

For $-99 < x < 0$, $\frac{121}{x}$ is more negative than $-\frac{11}{9}$, meaning that $\frac{121}{x}$ is farther from zero.

(9) $\boxed{-6 < x < 0}$ or $\boxed{x > 6}$; equivalent to $\boxed{x \in (-6, 0) \cup (6, \infty)}$. Notes: It would be **incorrect** to write that $|x|$ is greater than 6. You must treat the two cases separately:

- If $x > 0$, $\frac{x}{9} > \frac{4}{x}$ becomes $x^2 > 36$, for which $x > 6$.

- If $x < 0$, it becomes $x^2 < 36$ since the inequality reverses direction when multiplying both sides by a negative number, for which $|x| < 6$. Since x is negative in this case, $|x| < 6$ requires $x > -6$. For example, when x equals -5, observe that $-5 > -6$ (agreeing with $x > -6$), $|-5| < 6$ (agreeing with $|x| < 6$), $(-5)^2 < 36$ (agreeing with $x^2 < 36$), and $\frac{-5}{9} > \frac{4}{-5}$ (agreeing with $\frac{x}{9} > \frac{4}{x}$).

Calculator check:

- $\frac{-5.9}{9} \approx -0.6556$ is greater than $\frac{4}{-5.9} \approx -0.6780$ (since -0.6556 is less negative) and $\frac{-6.1}{9} \approx -0.6778$ isn't greater than $\frac{4}{-6.1} \approx -0.6558$ (since -0.6778 is more negative), which agrees with $-6 < x < 0$.

- $\frac{6.1}{9} \approx 0.6778$ is greater than $\frac{4}{6.1} \approx 0.6558$ and $\frac{5.9}{9} \approx 0.6556$ isn't greater than $\frac{4}{5.9} \approx 0.6780$, which agrees with $x > 6$.

(10) $\boxed{x > 0}$; equivalent to $\boxed{x \in (0, \infty)}$. Notes: Any positive value of x will satisfy $-\frac{27}{x} < \frac{x}{3}$. Here x can't be negative because the left-hand side would be positive while the right-hand side would be negative, and a positive number can't be less than a negative number. (As to the question of whether or not x could be zero, the left-hand side would be undefined.)

WAS THIS BOOK HELPFUL?

Much effort and thought were put into this book, such as:
- Covering a variety of intermediate algebra skills.
- Concisely introducing the main ideas at the beginning of each chapter.
- Solving examples step by step to serve as a helpful guide.
- Including checks or notes along with the answers at the back of the book.

If you appreciate the effort that went into making this book possible, there is a simple way that you could show it:

Please take a moment to post an honest review.

For example, you can review this book at Amazon.com or Goodreads.com.

Even a short review can be helpful and will be much appreciated. If you are not sure what to write, following are a few ideas, though it is best to describe what is important to you.
- Was it helpful to have checks/notes in addition to the answers at the back of the book?
- Were the examples useful?
- Were you able to understand the ideas at the beginning of the chapter?
- Did this book offer good practice for you?
- Would you recommend this book to others? If so, why?

Do you believe that you found a mistake? Please email the author, Chris McMullen, at greekphysics@yahoo.com to ask about it. One of two things will happen:
- You might discover that it wasn't a mistake after all and learn why.
- You might be right, in which case the author will be grateful and future readers will benefit from the correction. Everyone is human.

ABOUT THE AUTHOR

Dr. Chris McMullen has over 20 years of experience teaching university physics in California, Oklahoma, Pennsylvania, and Louisiana. Dr. McMullen is also an author of math and science workbooks. Whether in the classroom or as a writer, Dr. McMullen loves sharing knowledge and the art of motivating and engaging students.

The author earned his Ph.D. in phenomenological high-energy physics (particle physics) from Oklahoma State University in 2002. Originally from California, Chris McMullen earned his Master's degree from California State University, Northridge, where his thesis was in the field of electron spin resonance.

As a physics teacher, Dr. McMullen observed that many students lack fluency in fundamental math skills. In an effort to help students of all ages and levels master basic math skills, he published a series of math workbooks on arithmetic, fractions, long division, word problems, algebra, geometry, trigonometry, logarithms, and calculus entitled *Improve Your Math Fluency*. Dr. McMullen has also published a variety of science books, including astronomy, chemistry, and physics workbooks.

Author, Chris McMullen, Ph.D.

Put your intermediate algebra skills to the test with…

Includes good variety and instructive solutions.

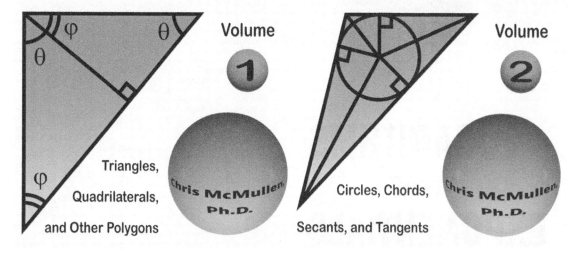

PLANE
GEOMETRY
Practice Workbook with Answers

PLANE
GEOMETRY
Practice Workbook with Answers

φ
θ
θ

θ

φ

Volume

1

Chris McMullen, Ph.D.

Triangles,

Quadrilaterals,

and Other Polygons

Volume

2

Chris McMullen, Ph.D.

Circles, Chords,

Secants, and Tangents

TRIGONOMETRY
Essentials Practice Workbook with Answers
Master Basic Trig Skills

$$\cos(\alpha + \beta) = \cos\alpha\cos\beta - \sin\alpha\sin\beta$$

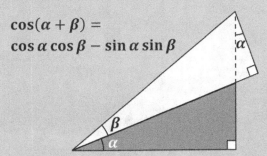

Improve Your Math Fluency Series

Chris McMullen, Ph.D.

TRIG IDENTITIES
Practice Workbook with Answers

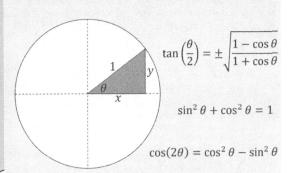

$$\tan\left(\frac{\theta}{2}\right) = \pm\sqrt{\frac{1 - \cos\theta}{1 + \cos\theta}}$$

$$\sin^2\theta + \cos^2\theta = 1$$

$$\cos(2\theta) = \cos^2\theta - \sin^2\theta$$

Chris McMullen, Ph.D.

Chris McMullen, Ph.D.

$\log_3 81 = ?$

$\log_4 32 = ?$

$\ln e = ?$

$e^0 = ?$

$\ln(e^x) = ?$

Essential   Skills   Practice   Workbook with Answers

$2^9 = ?$

$\cosh(x - y) = ?$

LOGARITHMS
and
EXPONENTIALS

Essential
CALCULUS
Skills Practice Workbook
with Full Solutions

$$\frac{d}{dx}\tan(5x)$$

$$\int \sqrt{1 - x^2}\, dx$$

Chris McMullen, Ph.D.

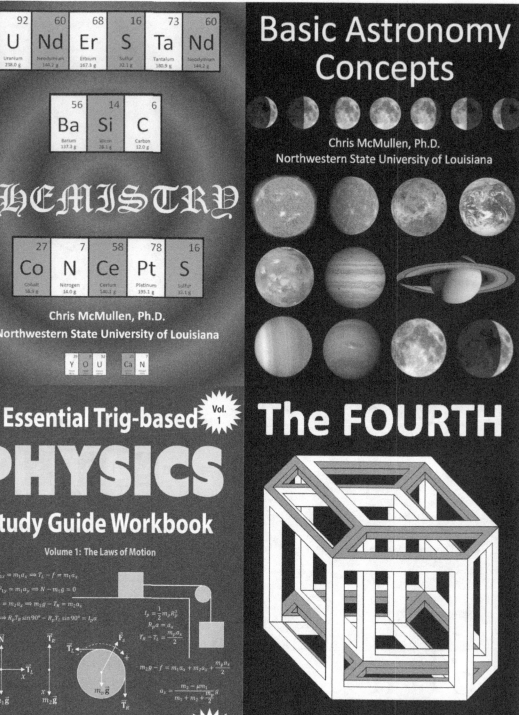

Made in the USA
Middletown, DE
03 June 2024

55217244R00130